Elementary Algebra

A Holistic Course

Antarctica

Gail D. Davis, PhD

Master IT! Academics

Elementary Algebra

A Holistic Course

COPYRIGHT @ 2008 Gail D. Davis
Master It! Academics

ALL RIGHTS RESERVED. No part of this work covered by the copyright hereon may be reproduced or used in any form or by any means – graphic, electronic, or mechanical, including photocopying, recording, or taping, Web distribution, or information storage and retrieval systems – without prior written permission of the publisher.

Printed in the United States of America

Regarding Mathematics, S Gudder said it this way:

The essence of mathematics is not to make things complicated, but to make complicated things simple.

My Dad said it this way:
Look deep enough into any topic and at its core is mathematics.

The Basic Tools of Mathematics

Workshop 1: Signed Numbers

In the real world, it is common for us to use words and phrases like debt, asset, money earned, money owed and temperature in everyday conversation.

Although you may not associate these everyday concepts with algebra, you will see in Workshop 1 that these and other real world situations often require the use of both positive (+) numbers and (-) negative numbers to accurately describe aspects of the world we live in.

☑ **Positive Numbers are greater than zero**

☑ **Negative Numbers are less than zero**

Figure 1.1

Figure 1.1 at the left shows an outdoor thermometer. The larger numbers on the outer scale range from a temperature of $+120^0$ F to (-60^0) F. Depending on how cold the air is, the temperature can be expressed as either a positive value (above zero) or a negative value (below zero) in degrees Fahrenheit.

The inner scale on the same outdoor thermometer includes smaller numbers that represent the air temperature read in degrees Celsius. Notice that the inner scale ranges from $+50^0$ Celsius to -50^0 Celsius.

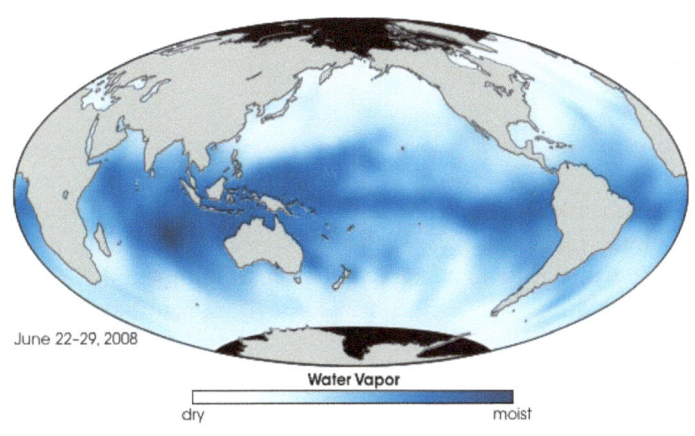

Table of Contents	*Page*
Workshop 1: Signed Numbers	
Investigation 1.1—Signed Numbers in the Real World	6
Investigation 1.2—The Number Line	21
Investigation 1.3—Adding Integers Using a Number Line	33
Investigation 1.4—Adding Signed Numbers	42
Investigation 1.5—Calculating Change Using Signed Numbers	53
Investigation 1.6—Finding Perimeter Using Algebraic Expressions	65
Investigation 1.7—Multiplying Signed Numbers	74
Investigation 1.8—Combining Concepts	88
Workshop 2: Using Fractions	
Investigation 2.1—Visual Fractions	100
Investigation 2.2—Multiplying Fractions	113
Investigation 2.3—Equivalent Fractions	124
Investigation 2.4—Adding Fractions	135
Investigation 2.5—Complex Fractions Using Division	145
Investigation 2.6—Calculating Simple Interest	153
Investigation 2.7—Solving Equations Using Fractions	158

Math Toolbox

Workshop 1: Signed Numbers

Investigation 1.1—Signed Numbers in the Real World

We will begin our holistic look at mathematics by considering both positive and negative numbers. Many times positive numbers alone are not sufficient for representing many situations in the real world. For example to express debt, the depth above or below sea level, and temperature (to name just a few) requires the use of numbers that are both less than and greater than zero. Investigation 1.1 will focus on applying the correct sign, either (+) or (-) to quantities found in the real-world that often need to be described as being positive or negative.

Note: We will assume that sea level is represented as the "zero" point so that water surface elevations *above* zero are positive, and water surface elevations *below* zero are described using a negative value.

Example 1: Figure 1.2 shows the Caspian Sea situated between Russia and Kazakhstan. In addition to being the largest enclosed body of water on Earth, its surface area extends 143,200 square miles and its surface elevation lies 28 meters below sea level. 28 meters is about the same distance as 92 feet.

Instructions: Using only numbers and the correct sign, answer the following questions:

1. How much area in square miles does the Caspian sea cover?
   ⇒ **Answer: +143,200**

2. What is the surface elevation of the Caspian sea in meters?
   ⇒ **Answer: −28 meters (28 meters below sea level)**

3. What is the surface elevation of the Caspian sea in feet?
   ⇒ **Answer: -92 feet (92 meters below sea level).**

Figure 1.2

Example 2: The card game, Canasta is believed to have originated in Montevideo, Uruguay in 1939 and spread from there to the United States. Canasta was very population in the U.S. in the 1950's. In Canasta, it is possible to end up with a negative score by losing more points than one has earned.

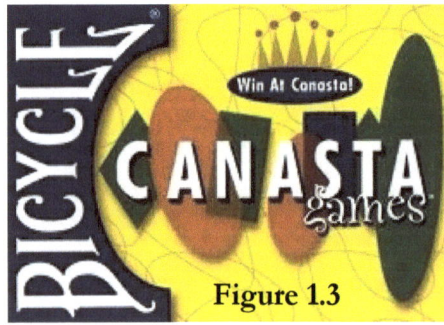

Figure 1.3

Instructions: Using only numbers and the correct sign, answer the following questions:

1. What is the new score in a game of Canasta if a player loses 10 points from a beginning score of zero?
   ⇒ **Answer: -10 (read "negative" ten points).**

2. A player begins a hand of Canasta with 20 points. What is his new score if he loses 25 points?
   ⇒ **Answer: -5 points (read "negative" 5 points).**

Math Toolbox

Workshop 1: Signed Numbers

Investigation 1.1—Signed Numbers in the Real World

Example 3: In 1905 the mines of Park City, Utah produced more precious metals than all the mining camps in neighboring states. Between 1880 and 1920 Park City mines produced more than 20 million tons of ore including lead, zinc, copper, silver and a small amount of gold.

In 1903 the miners would often work for 10 hours each day, six days a week deep inside the mine below the Earth's surface. Their average pay for a days work was only $3.00 Figure 1.4 shows the remnant of an old mine in Park City Utah. (Photos courtesy of the Park City Museum).

Note: Assume that "zero" represents the surface of the Earth and depths below the surface are described as "negative" values.

1. A miner finds himself 100 feet below the surface of the Earth. Represent the miner's position below the Surface as a number.

    ⇒ **Answer: -100 feet**

2. A miner begins his day at a depth of 50 feet below the surface of the Earth. By the end of the 10 hour day, he traveled another 20 feet deeper into the mine. At the end of the day, what was the miner's position below the Earth?

    ⇒ **Answer: -70 feet**

3. A miner worked for 6 days each week. Assuming his average pay was $3.00 per day, how much would he earn for the week? (Express your answer as a signed number).

    ⇒ **Answer: +$18.00**

4. The Cage shown in Figure 1.6 is a mechanical device for transporting the miners underground, sometimes as deep as 300 feet below the Earth's surface. Express this distance as a signed number.

    ⇒ **Answer: -300 feet**

5. Assuming that a miner is working 275 feet underground. How far up from that position would he have to travel in the cage so that his position would be only 175 feet below ground? Express your answer as a signed number.

    ⇒ **Answer: +100 feet.**

Figure 1.4

Figure 1.5

Figure 1.6

Investigation 1.1—Signed Numbers in the Real World

Workshop 1: Signed Numbers

Example 4: Temperatures are another example of real-world concepts that sometimes need to be expressed using both positive and negative numbers. In this example, we will investigate the Earth's southernmost continent that overlies the South Pole. On average, Antarctica is by far the coldest, driest, and windiest continent on Earth and has the highest average elevation of all continental land masses.

Let's learn a little about each of the three stations depicted in Figure 1.7.

1. **McMurdo station** is a coastal station whose weather is influenced by the South Pacific Ocean. During the winter as the sun disappears below the horizon the temperatures plummet. McMurdo station is located at a height of 24 meters (about 78 feet) above sea level and sees temperatures ranging from -3^0 C to a blistering -26^0 C.

2. The temperatures at the **Mundsen-Scott South Pole station** vary on 6 month cycles according to the position of the sun and whether or not it remains permanently dark. When permanently dark, the temperature is very stable and cold, but as soon as the sun begins to rise higher and higher in the sky and the days get longer, the temperature begins to increase. The temperature at the South Pole station varies from about -28^0 Celsius in December to a balmy -60^0 C in August.

Figure 1.7. The Antarctic peninsula is home to scientists from 27 different countries that conduct experimental research not reproducible any other place on Earth. The graphic indicates just three of the most popular research stations: McMurdo (the most populated), Vostok, and the South Pole.

3. **Vostok station, a Russian research station established in the Antarctic December 16, 1957,** is the Earth's Southern Geomagnetic Pole. Although it was temporarily closed in 1994, it has operated full-time for the last 37 years as a cooperative effort among scientists from the United States, Russia and France. July 21, 1983 the coldest temperature ever recorded anywhere on Earth, a frigid -89.2^0 C, was measured at this site. On the Fahrenheit scale the corresponding temperature is -128.6^0 F.

To aid the research efforts at each of these three sites, a precise record of daily temperatures and rainfall equivalent is documented. Because actual liquid rain is rare in Antarctica, the average monthly precipitation measure is given as a "rainfall equivalent" - *the amount of precipitation that would have fallen as rain and not snow.* With this daily record, scientists have been able to produce averages, by month, over a year's time for four important temperature and moisture variations: (i) the average daily temperature in ^{0}C, (ii) the average daily maximum temperature, (iii) the average daily minimum temperature and (iv) the average monthly rainfall in millimeters.

Many times data tables, such as the one shown on the next page for average daily temperature by month, are cumbersome and difficult to use. Scientists often graph the data and connect each point with a line graph so that the data collection can tell an entire story visually by noting the overall relationship among the points. This relationship is one of the tools in mathematics that crosses nearly every discipline: science, psychology, education, business, history, exercise science and kinesiology, and even music, to name a few. In this example you will be using a graph to identify points and place them in a data table. Then we reverse the activity. You will be given a data table and asked to enter the points in the correct position on a graph.

Math Toolbox

Workshop 1: Signed Numbers

Investigation 1.1—Signed Numbers in the Real World

Average Daily Temperature (°C) for McMurdo Station, Antarctica.

Month	Jan	Feb	Mar	Apr	May	Jun	Jul	Aug	Sept	Oct	Nov	Dec
Temperature (°C)	-2.9	-9.5	-18.2	-20.7	-21.7	-23	-25.7	-26.1	-24.6	-28.9	-9.7	-3.4

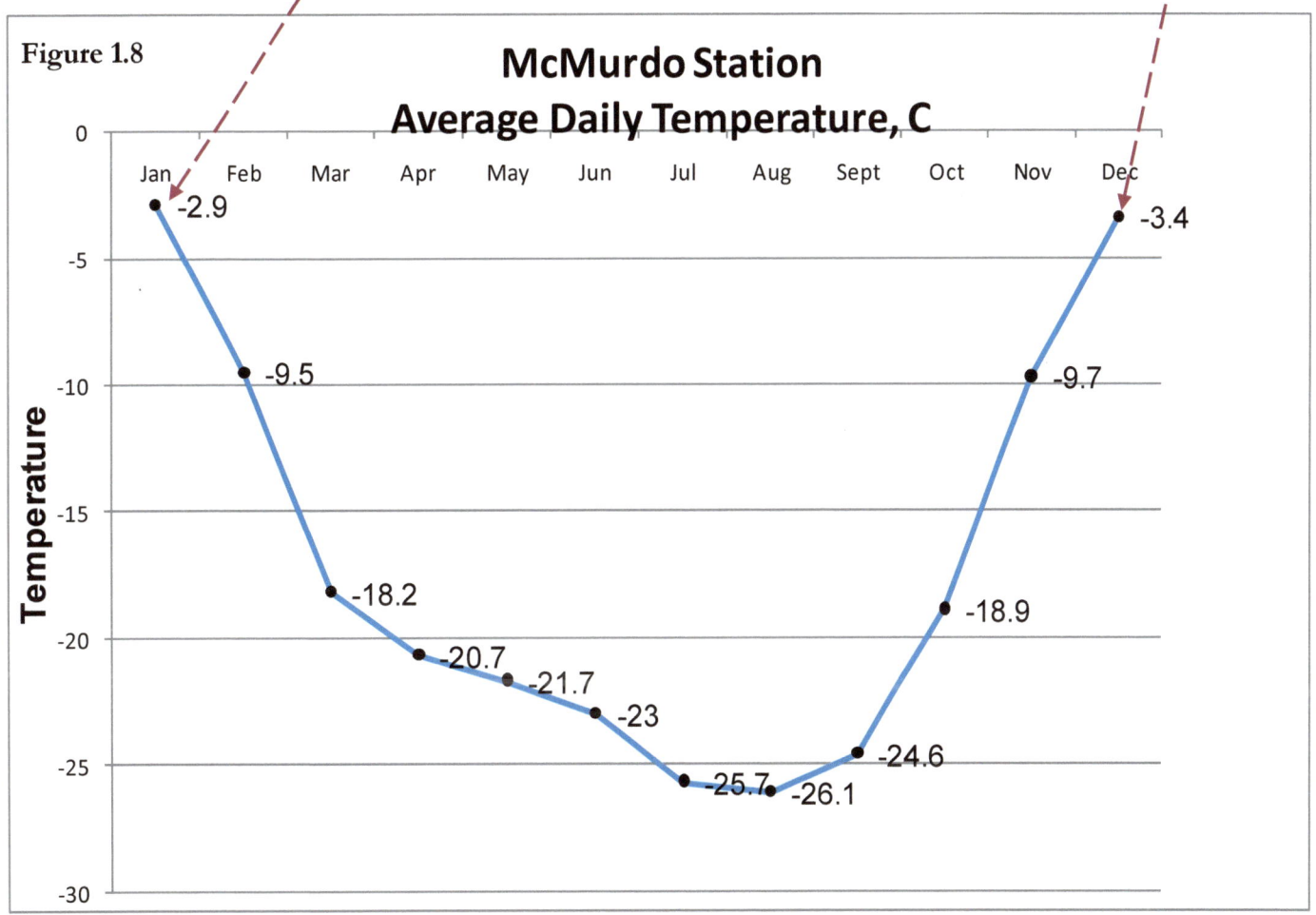

Figure 1.8

By using the graph above, answers the questions below regarding the overall pattern that the data depicts.

1. Describe the shape of the graph. What does this tell you about the changing temperatures from January through December?

⇒ **Answer:** Beginning in December, the graph shows a downward trend in temperature, with the lowest point associated with the month of August at –26.1°. After reaching this low point, the average monthly temperatures begin to increase ending at –3.4° C in December.

2. What conclusion can you draw about the "warm" months in Antarctica? Are they similar to the pattern of summer months in North America?

⇒ **Answer:** According to the graph, "summer" in Antarctica is in December and January; Winter is found April—October, with the coldest months shown in June, July & August. This is opposite of summer and winter in North America.

Investigation 1.1—Signed Numbers in the Real World

Workshop 1: Signed Numbers

Example 5: The last example that we will investigate is one that you no doubt understand very well. Because a bank account is considered indispensable by most individuals and businesses in today's society, you will have some first hand experience with this example. Briefly, a bank acts as a payment agent by maintaining checking and savings accounts for customers, paying checks drawn by customers on the particular bank in question, and collecting the funds from checks deposited to customer accounts.

Most of us have probably had the misfortune of being overdrawn at least once in our life and have dealt with the dreaded "negative balance". If you haven't as yet, your time will surely come! In banking we use two terms that relate to the study of positive and negative numbers.

A ***debit*** is an amount subtracted from your account. Hence, a debit is a "negative" value because it causes your balance to decrease.

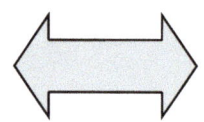

A ***credit*** is an amount added to your account. Hence, a credit is a positive amount because it causes your balance to increase.

Joan opens a checking account on the first day of the month with a starting balance of $1,000.00.

During the first month, Joan writes the following checks or authorizes automatic withdrawals from her checking account in the following amounts. The table summarizes the debits presented to the bank for payment out of Joan's account.

Day	Debit Amount	Day	Debit Amount
2	-140.00	15	-30.00
4	-58.00	19	-25.00
8	-10.00	28	-120.00

What is the total *debit* amount?

Adding up the individual debit values we see that the total debits charged to Joan's account is –383.00.

During the same month, Joan's account is credited with the following amounts from her salary and sales commissions.

Day	Credit Amount	Day	Credit Amount
1	+1,000.00	17	+250.00
3	+150.00	20	+35.00
7	+85.00	29	+250.00

What is the total *credit* amount?

Adding up the individual credits we see that the total money "added" to Joan's account (including her initial deposit of $1,000.00) is +1,770.00

To find Joan's balance we add the debit amount to the credit amount.

$$+1770.00 + (-383) = 1770.00 - 383 = +1387$$

When "adding" a debit, we just subtract. Adding a negative number results in subtraction.

Workshop 1: Signed Numbers

Investigation 1.1—Signed Numbers in the Real World

💰 Joan's checking account after the first month has a balance of (+$1387.00). The graph below shows the daily increase and decrease in Joan's account as credits are added and debits are subtracted.

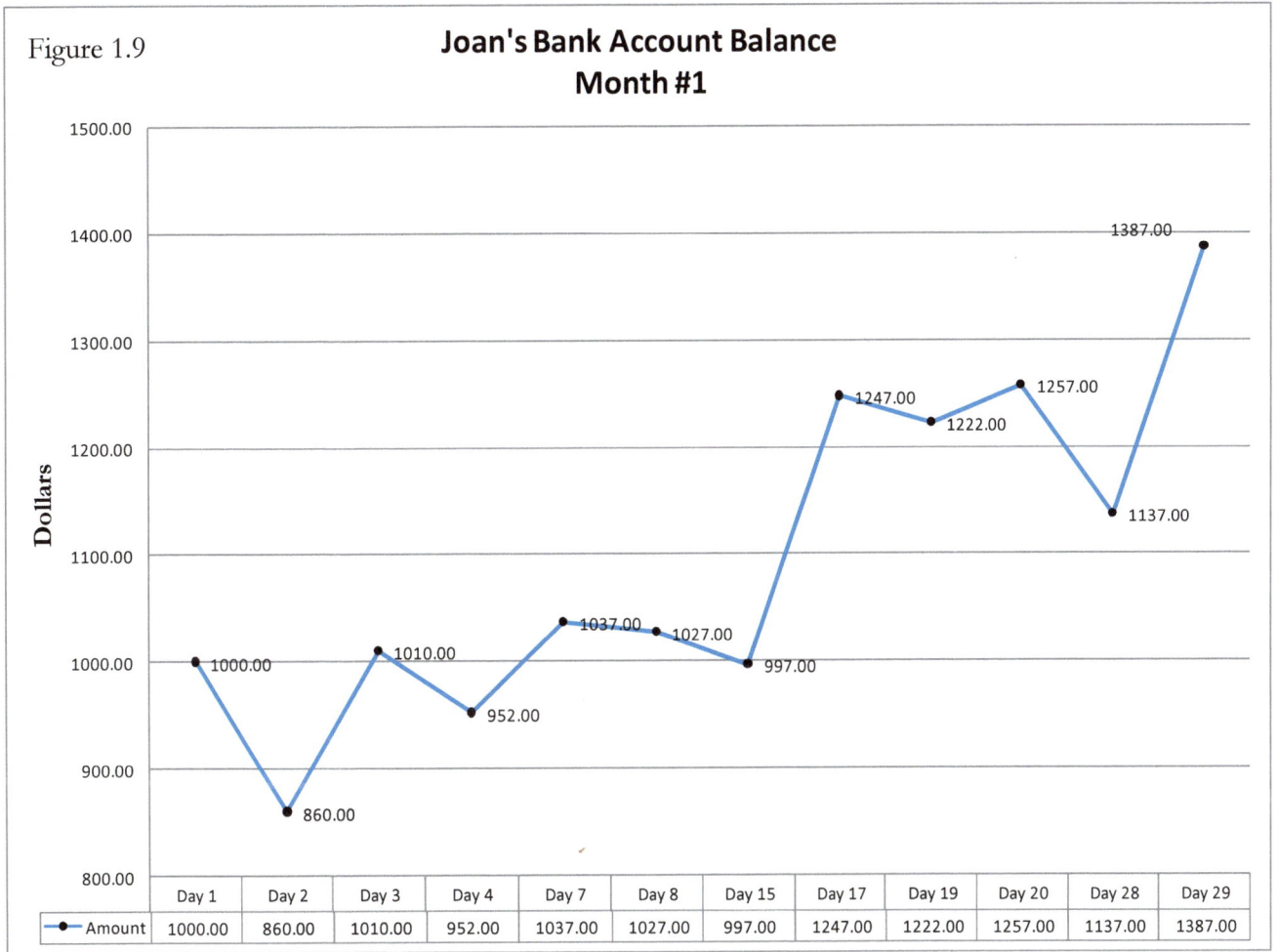

Figure 1.9

Using the information in the graph and table above to answer these questions:

1. By how much did Joan's bank account change between Day 1 and Day 2?
   ⇒ Joan's balance decreased from +$1,000.00 to +$860.00. This decrease occurred because a "debit" of -$140.00 was added to Joan's account. Notice the line between Day 1 and Day 2 is slanted downward. This is an indication that Joan's bank balance decreased.

2. By how much did Joan's bank account change between Day 4 and Day 7?
   ⇒ Joan's balance increased from +$952.00 to +$1037.00. This increase occurred because a "credit" of +$85.00 was added to Joan's account. Notice the line between Day 4 and Day 7 is slanted upward. This is an indication that Joan's bank balance increased.

Math Toolbox
Workshop 1: Signed Numbers
Investigation 1.1—Signed Numbers in the Real World

Using the information provided by the graph and table on the previous page to answer these questions:

3. **By how much did Joan's bank account change between Day 7 and Day 8?**

⇒ Joan's balance decreased from +$1,037.00 to +1027.00 This decrease occurred because a "debit" of -$10.00 was added to Joan's account on Day 8.   Notice the line between Day 7 and Day 8 is slanted slightly downward showing a slight decrease

4. **By how much did Joan's bank account change between Day 15 and Day 17?**

⇒ Joan's balance increased from +$997.00 to +$1247.00. This increase occurred because a "credit" of +$250.00 was added to Joan's account on Day 17.   Notice the line between Day 15 and Day 17 makes a strong slant upward. This is an indication that Joan's bank balance increased by quite a bit between these two days.

5. **By how much did Joan's bank account change between Day 17 and Day 19?**

⇒ Joan's balance decreased from +$1247.00 to +1222.00. This decrease occurred because a debit of -$25.00 was added to Joan's account on Day 19.

6. **By how much did Joan's bank account change between Day 19 and Day 20?**

⇒ Joan's balance increased from +$1222.00 to +$1257.00. This increase occurred because a credit of +$35.00 was added to Joan's account on Day 20. Notice the line is slanted up, indicating Joan's bank balance increased between these two days.

7. **By how much did Joan's bank account change between Day 20 and Day 28?**

⇒ Joan's balance decreased from +$1257.00 to +$1137.00. This decrease occurred because a debit of -$120.00 was added to Joan's account on Day 28. Notice the line is slanting downward between these two days indicating a significant decrease in her balance.

8. **By how much did Joan's bank account change between Day 28 and Day 29?**

⇒ Joan's balance increased from +$1137.00 to +$1387.00. This increase occurred because a credit of +$250.00 was added to Joan's account on Day 29. Notice the sharp increase in the line between these two days indicating a significant increase in her balance.

✎ Assignment #1　　☑ Investigation 1.1　　Name

Instructions: Read each of the following practice exercise paragraphs. Then, answer the questions posed and be sure to provide your answer using the correct signed number.

I. Death Valley California is often referred to as the hottest, driest and lowest place in North America. This is because its surface elevation is the lowest in the United States measuring 282 feet below sea level. The highest mountain at Death Valley National Park is 11,049 feet above sea level and is called Telescope Peak. There is one interesting historical story associated with Death Valley—how it got its name. As the story goes, in the winter of 1849—1850, a group of pioneers became lost and one of their group died. After this, they all assumed *Death Valley* would be their final resting place. But that was not to be their fate because the group was soon rescued by two scouts, William Manly and John Rogers. As the pioneers and two scouts climbed out of the valley over the Panamint Mountains, one of the men turned and looked back to say, "Goodbye Death Valley". The name and the legend of the lost 49er's has stuck ever since.

Figure 1.10. The Panamint Mountains

Figure 1.11: Furnace Creek Inn

The highest *ground* temperature in Death Valley was recorded at Furnace Creek on July 15, 1972—a whopping 201^0 F. Similarly, the maximum air temperature on that same day was recorded at 128^0 F. In addition to its lowest point of 282 feet below sea level, the area around Death Valley also has higher elevations with cooler temperatures than those in the lower valley. At the higher elevations, the temperatures drop between 3^0 and 5^0 every 1000 feet above the ground.

1. What is the highest ground temperature recorded at Death Valley?
2. How high is Telescope Peak?
3. In feet, what is the lowest point in Death Valley?
4. Assuming the 49'ers were rescued at the lowest point at Death Valley, how high did they have to climb to reach a height of 950 feet above sea level in the Panamint Mountains?
5. At the higher elevations how much do the temperatures drop for each 1,000 feet above ground?

Challenge Question 1: Assume that the temperature for a particular area in Death Valley located at sea level dropped 4^0 F for every 1,000 feet above sea level. How much would the temperature have decreased at an elevation of 3,000 feet above sea level? Write this temperature decrease as a signed number.

Challenge Question 2: Based on your answer for Challenge Question 1 above, if the temperature at sea level was 5^0 F, what would the temperature be at 3,000 feet? (Write your answer as a signed number).

Assignment #2
Page 1 of 2   ☑ Investigation 1.1   Name

II. **The Mariana Trench,** a deep depression in the floor of the Pacific Ocean, is located near Japan just east of the 14 Mariana Islands. This trench is one of just 22 major trenches in the world and is the deepest seafloor depression on Earth. At a point about 340 miles southwest of Guam (about 210 miles) is the lowest point on Earth. Called the *Challenger Deep,* its depth is estimated to be 36,198 feet below the surface of the ocean (11,033 meters). Of the four Oceans on Earth—the Arctic Ocean, The Atlantic Ocean, the Indian Ocean, and the Pacific Ocean—their average depths below sea level vary along with the deepest point in each.

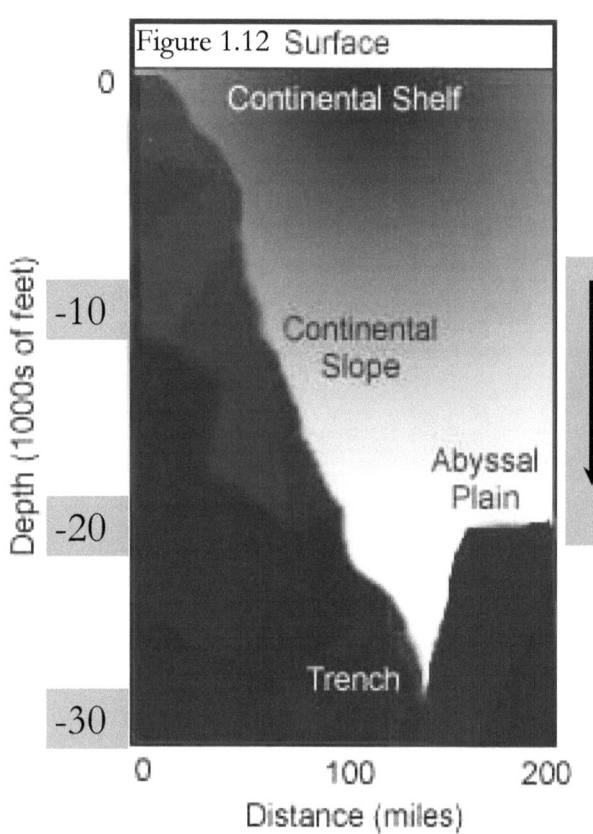

- The average depth of the Indian Ocean is 3,872 meters or about 12,740 feet below sea level. The deepest point in the Indian Ocean is the Java Trench located 7,725 meters below the ocean's surface (or about 25,344 feet deep).

- The average depth of the Atlantic Ocean is 3,926 meters or about 12,881 feet below sea level. The deepest point in the Atlantic Ocean is located at the Puerto Rico Trench 8,648 meters below the ocean's surface. (About 28,374 feet deep).

- The average depth of the Arctic Ocean is 1,038 meters corresponding to about 3,407 feet deep. The lowest point in the Arctic Ocean is the Eurasian Basin that lies about 5,450 meters below the surface (17,881 ft deep).

- The average depth of the Pacific Ocean 4,188 meters, or about 13,740 feet below sea level.

The table below organizes each of the 4 oceans according to (i) their average depth below sea level and (ii) the deepest point in each. Using the correct signed numbers, fill in the missing information.

Ocean Name	Average Depth below Sea Level (meters)	Average Depth Below sea Level (feet)	The Deepest Point (meters)	The Deepest Point (feet)
Atlantic Ocean				
Arctic Ocean				
Pacific Ocean				
Indian Ocean				

Assignment #2
Investigation 1.1

Using the information on the previous page, answer the following questions. Write your answer as a signed number.

(a) **Use Figure 1.12.** What ocean depth lies between 10,000 feet below the ocean's surface and 20,000 feet below the ocean's surface?

(b) **Use Figure 1.12.** What depth does (-30) refer to along the vertical axis of the picture?

(c) A SCUBA diver is motionless at a depth of 35 feet below the Ocean's surface. How far must she travel to reach a depth of 55 feet?

(d) On January 23, 1960 Jacques Piccard, a Swiss Ocean Engineer, and Donald Walsh a U.S. Naval Lieutenant descended to the deepest point in the Pacific Ocean in a French built submarine specifically designed to operate at great depths, setting a world record. They found the bottom of the Mariana Trench at a distance below the ocean's surface of about 35,810 feet. Assuming the two explorers are at the bottom of the ocean at this depth, how many feet do they need to ascend in their submarine to be just 5,010 feet below the ocean's surface?
(Hint: Draw a picture)

(e) A submarine enters a deep part of the Atlantic Ocean and descends to a depth of 400 feet below the surface. Then, the submarine ascends toward the surface traveling a distance of 55 feet. After reaching this new position, the submarine is ordered to once again descend 125 feet down from its current position. How far from the ocean's surface does the submarine end up? (Hint: Use the Schematic below to help you visualize the problem).

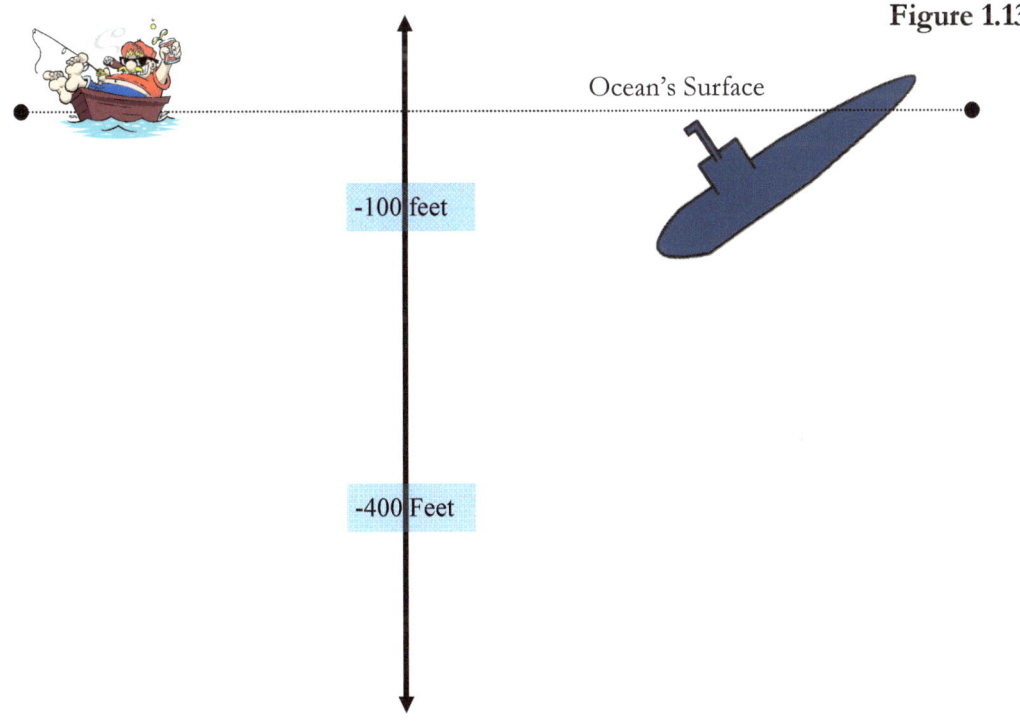

Figure 1.13

✎ Assignment #3   ☑ Investigation 1.1   Name

III. The temperatures each day in Antarctica vary considerably due to the extreme positions of the sun and the energy (or lack thereof) emitted to this part of the globe. Figure 1.14 below is a graph of the average daily "maximum" temperatures in °C collected for the McMurdo Research Station. As researchers recorded the daily temperatures each month, (over many cycles of collected data for each month), an average "maximum" daily temperature for the indicated month was calculated. For instance, the average "highest" temperature in January was calculated to be –25.9° C. The graph below is the result of that work. Use this information to fill in the missing cells of the temperature table below.

Average Daily Maximum Temperature (° C) for McMurdo Station, Antarctica.

Month	Jan	Feb	Mar	Apr	May	Jun	Jul	Aug	Sept	Oct	Nov	Dec
Temperature												

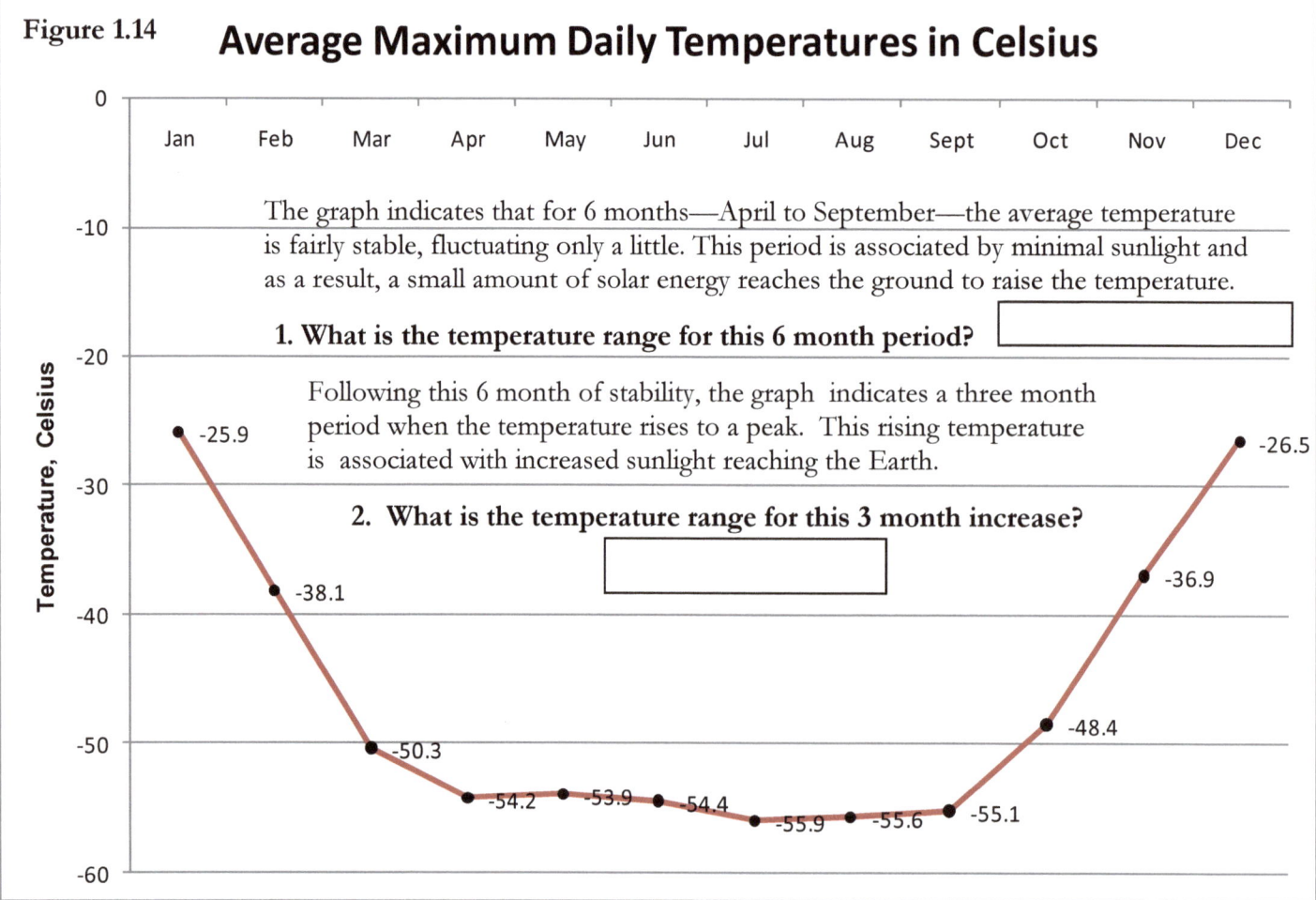

Figure 1.14 **Average Maximum Daily Temperatures in Celsius**

The graph indicates that for 6 months—April to September—the average temperature is fairly stable, fluctuating only a little. This period is associated by minimal sunlight and as a result, a small amount of solar energy reaches the ground to raise the temperature.

1. What is the temperature range for this 6 month period?

Following this 6 month of stability, the graph indicates a three month period when the temperature rises to a peak. This rising temperature is associated with increased sunlight reaching the Earth.

2. What is the temperature range for this 3 month increase?

Assignment #4 — Investigation 1.1 — Workshop 1

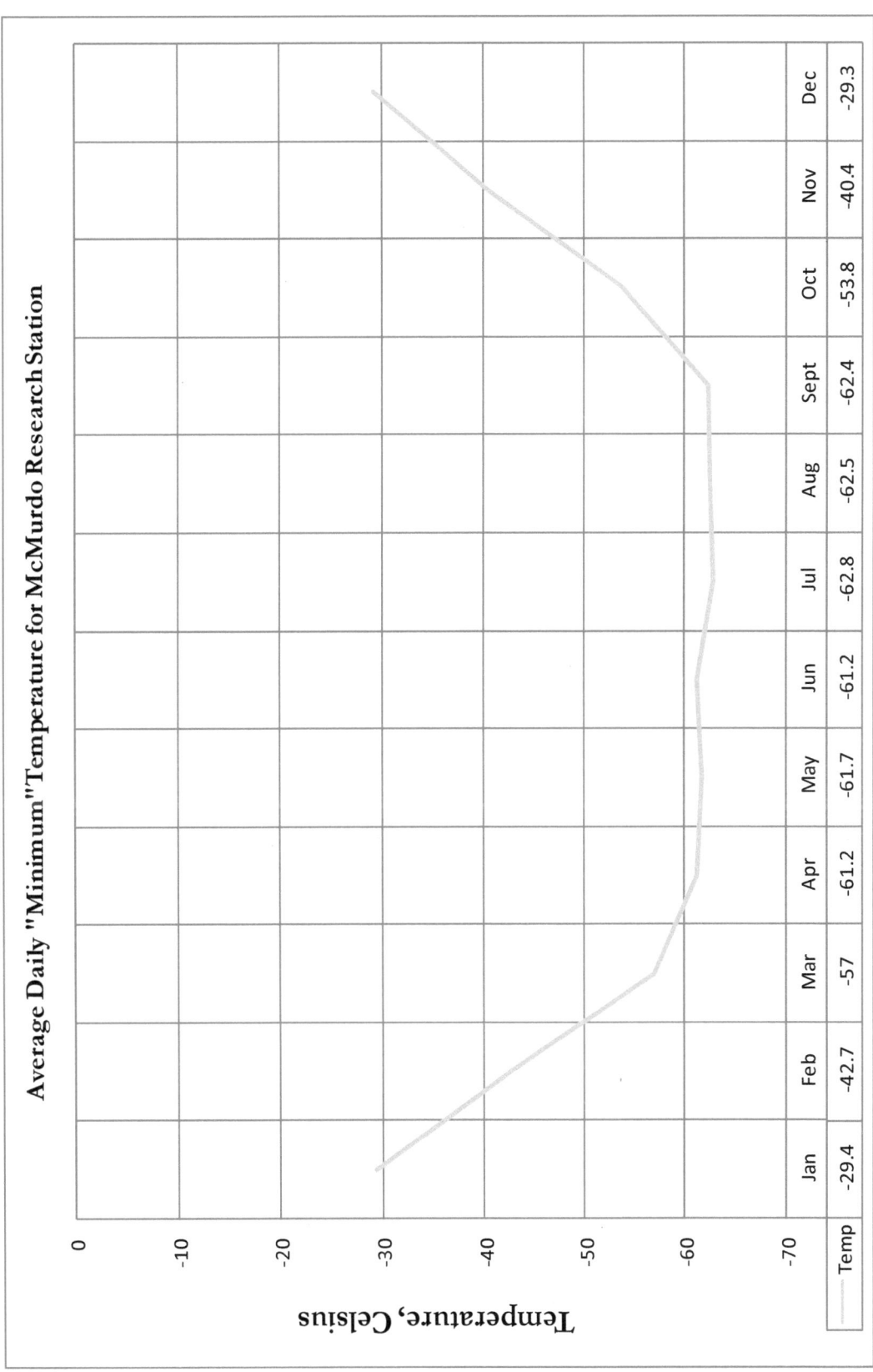

Figure 1.15

Using the temperature data at the bottom of this chart, locate the right spot on the grid and color in a small circle showing where each of the points should be located. A Grey line graph is shown to help you. Hint: Each of the temperature points will be drawn somewhere on the grey line.

Average Daily "Minimum" Temperature for McMurdo Research Station

	Jan	Feb	Mar	Apr	May	Jun	Jul	Aug	Sept	Oct	Nov	Dec
Temp	-29.4	-42.7	-57	-61.2	-61.7	-61.2	-62.8	-62.5	-62.4	-53.8	-40.4	-29.3

Assignment #5 — Investigation 1.1 — Workshop 1

Using the average rainfall data at the bottom of this chart, locate the right spot on the grid and color in a small circle showing where each of the points should be located. A Grey line graph is shown to help you. Hint: Each of the rainfall average points will be drawn somewhere on the grey line.

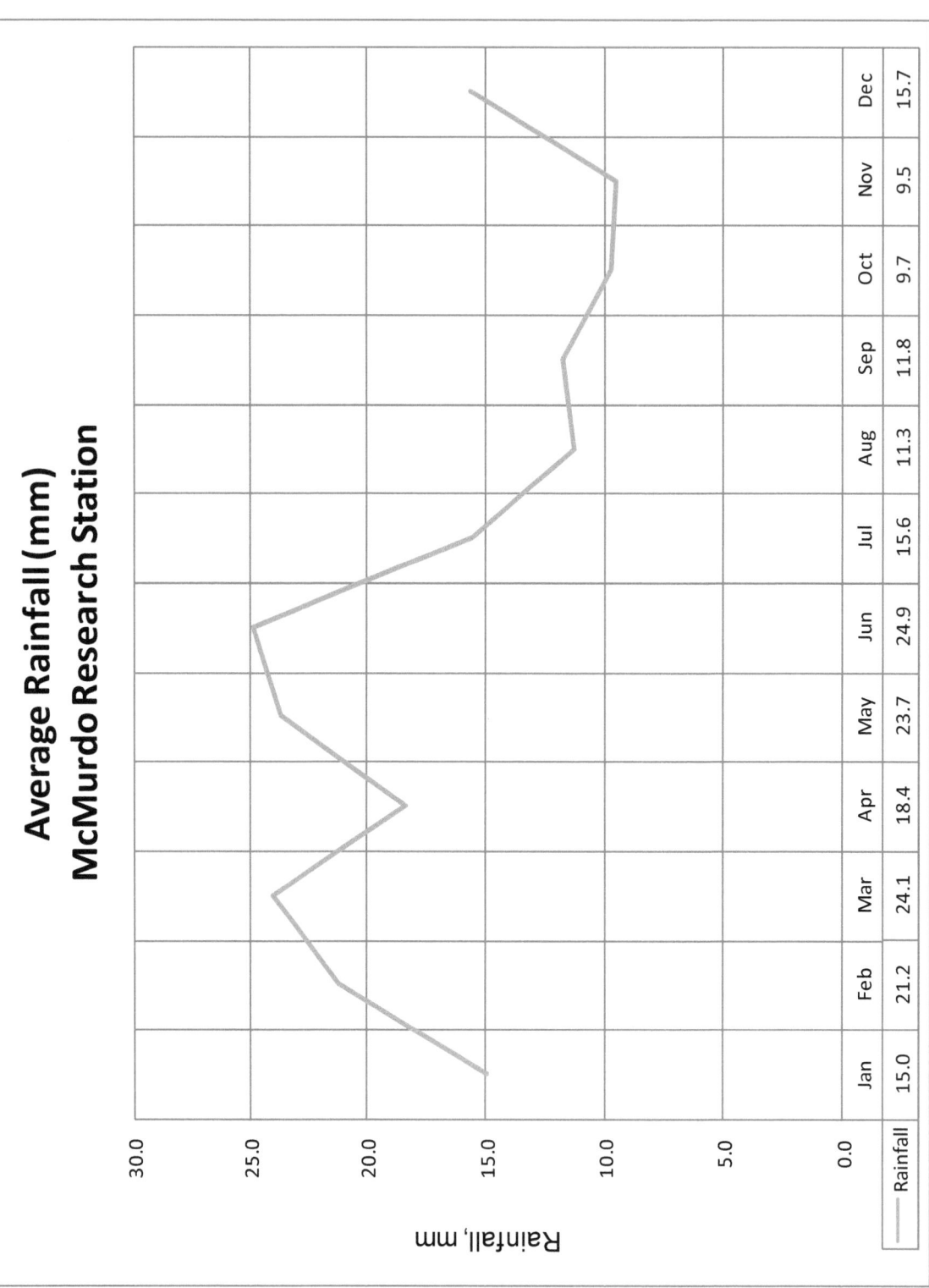

Figure 1.16

Average Rainfall (mm)
McMurdo Research Station

Rainfall	Jan	Feb	Mar	Apr	May	Jun	Jul	Aug	Sep	Oct	Nov	Dec
	15.0	21.2	24.1	18.4	23.7	24.9	15.6	11.3	11.8	9.7	9.5	15.7

✎ Assignment #6   ☑ Investigation 1.1   Name

IV. After the first month, Joan had a bank balance of +$1387.00. During Month #2, Joan graphed each credit and each debit added to her account. The graph below shows the result of Joan's effort during Month #2. Use this graph to enter the appropriate *Debit Amount* and *Credit Amount* in the table added each day to Joan's account.

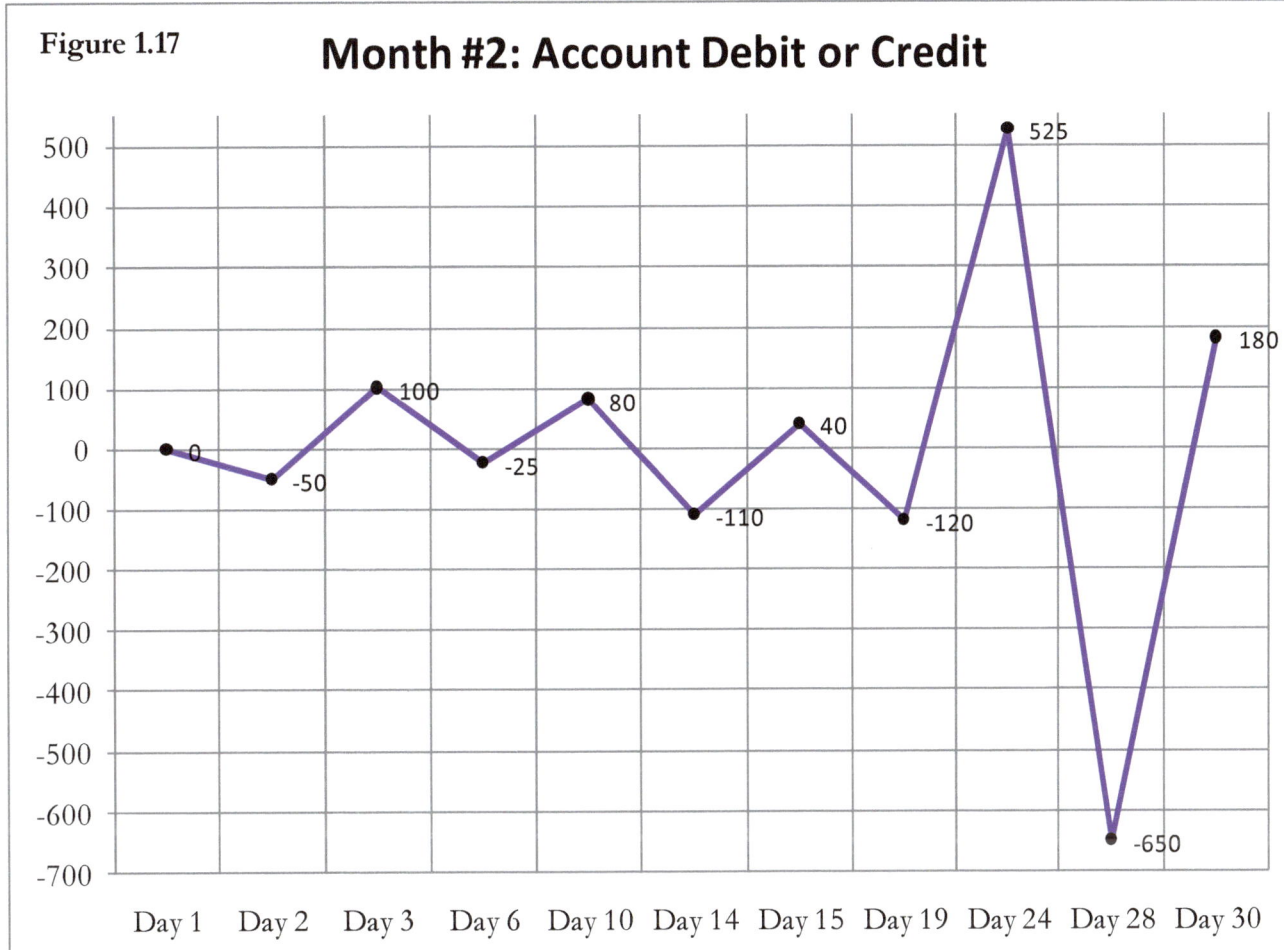

Figure 1.17 — Month #2: Account Debit or Credit

Day	Credit Amount	Debit Amount	Day	Credit Amount	Debit Amount
1			15		
2			19		
3			24		
6			28		
10			30		
14					

✎ Assignment #7   ☑ **Investigation 1.1**   Name

Using the completed table for Month #2, answer the following questions and don't forget to give your answer as a correct **signed number.**

1. Sum the total credits added to Joan's account. Enter the total in the box.

2. Sum the total debits added to Joan's account. Enter the total in the box.

3. Enter Joan's balance at the end of Month #1 in the box.

4. Enter the new balance at the end of Month #2 in the box.

5. When a debit (or negative value) is added, describe the line segment on the graph indicating this.

6. When a credit (or positive value) is added, describe the line segment on the graph indicating this.

7. Add. $(+200) + (-150) =$

8. Add. $(-200) + (-150) =$

9. In problem #7 you were asked to sum a credit and a debit. Explain what you were asked to do in problem #8.

☑ Skill Builder #1   Name _____

Instructions: Represent each quantity by an integer

1. A Scuba diver is swimming 42 feet below the surface of the water in the Gulf of Mexico.

2. At the end of the fiscal year in 1997, Apple Computer company reported an overall loss in revenue of $1045 million.
   (Source: Apple Computer, Inc).

3. Jack and Jim are getting certified to become Scuba divers. On a practice dive in the Florida Keys, Jack is at 45 feet below the surface, while Jim is 38 feet below the surface. Represent the depths for each diver by an integer. Determine who is deeper in the water.

4. The temperature during one February day in Fargo, North Dakota was $15°$ below zero on the Celsius scale.

   (a) Represent this temperature as a signed number.

   (b) Is this temperature colder or warmer than $20°$ below zero on the Celsius scale?

5. In 2004 an investment entrepreneur reported a 13% loss in revenue from the previous year. Express this loss by using an integer.

6. Wanda's Widgets, Inc. reported a net loss of $3,650 during the year 2007.

7. Joan's Checking account balance in January showed a credit of $350.00. In March her account balance showed a debit of $400.00. Express each of these account balances as positive or negative integers.

8. A submarine is at a depth 100 meters below the ocean's surface. The captain receives instructions to ascend toward the surface at a position 40 meters below the ocean's surface. How far up from his current position will the submarine ascend?

9. The temperature in Barrow, Alaska is $-25°$ F. By noon the temperature rose $5°$ F. What is the temperature at noon?

10. Explain why the number -335 is smaller than the number -235.

11. Mary gets a credit card statement in the mail. Her statement shows a balance of $55.00. Write this number as an integer.

12. The lowest point in the Johnson family swimming pool is 8 feet below ground level. Mary Johnson is only 4 feet tall. If she stands on the bottom of the pool, how much distance is there between her head and the surface?

13. How much colder is $-23°$ C than $-18°$ C?

☑ Skill Builder #2   Name _____

Instructions: Use the > sign to indicate that the integer on the left is greater than the integer on the right.
Example: 5 > 1
Use the < sign to indicate that the integer on the left is less than the integer on the right.
Example 6 < 25

1. 10   15

2. 25   12

3. 125   225

4. 20   0

5. 0   -10

6. -1   -3

7. -10   0

8. -5   -3

9. -100 +100

10. 10   -13

11. At the end of Month #1 Joan's bank account balance was $450. The next day Joan withdrew $525 from her account. Is her new balance > (greater than) or < (less than) zero?

   New Balance _____ Zero.

12. Tom and Ted are diving off the Florida Gulf Coast. Tom is 28 feet below the surface of the ocean. Ted is 10 feet below the surface of the ocean. Use the correct sign, < or >, to correctly indicate their relationship to each with respect to the surface.

   Ted's depth _____ Tom's Depth

13. Two miners report for work and are assigned different locations below the Earth. The first miner's location is 25 feet below the surface; the second miner's location is 13 feet below the surface. Use the correct sign, < or >, to correctly indicate their relationship to each other with respect to the surface.

   Miner #1 _____ Miner #2

14. Bob and Ben play 18 holes of Golf. At the end of play, Bob shoots 3 under par, while Ben shoots 4 over par. Use the correct sign, < or >, to correctly indicate their respective golf scores.

   Ben's Score _____ Bob's Score

15. During the first quarter, the local amateur football team lost 15 years on play #1 and gained 20 yards on play #2. During the second quarter the local amateur football team gained 25 yards on play #1 and lost 30 yards on play #2. Use the correct sign, < or >, to determine the net loss or gain in yards between quarter #1 and quarter #2.

   Quarter #1 _____ Quarter #2

Math Toolbox

Workshop 1: Signed Numbers

Investigation 1.2: The Number Line

Conceptualize means to arrive at a generalization as a result of things seen or previously experienced.

Investigation 1.1 introduced you to many real-world situations that required the use of both positive and negative numbers. Our goal in Investigation 1.2 is to expand the use of positive and negative numbers by learning to conceptualize the ideas introduced earlier.

Our study will now take a more theoretical turn, moving from specific applications in the real world to a conceptualization in the use of positive and negative numbers by using the **number line.**

A number line provides a one-dimensional picture of the ordered relationship between individual numbers and contains all <u>real numbers</u> continuing forever in both directions. A number line is useful when learning to add and subtract, especially when negative numbers are involved. The arrows indicate the number line goes on forever in both directions.

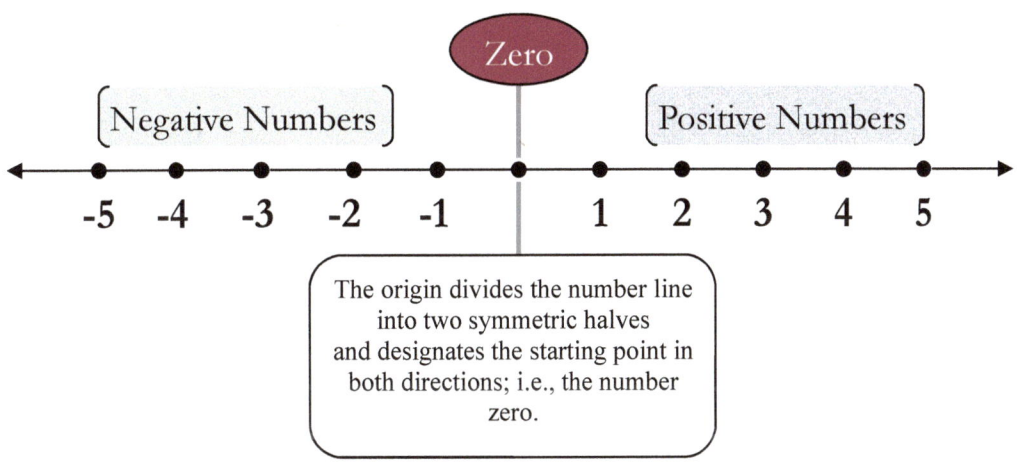

✷ For any two numbers graphed on the number line, the number to right is the greater number. To indicate symbolically that one number is greater than another we use the greater than sign, >
Example: **5 > 2 is read, "Five is greater than 2".**
*5 lies to the **right** of the number 2.*

◆ The number to the left is the smaller number. To indicate symbolically that one number is less than another we use the less than sign, <.
Example: **2 < 4 is read, "Two is less than 4"**
*2 lies to the **left** of the number 4*

Math Toolbox

Workshop 1: Signed Numbers

Investigation 1.2: The Number Line

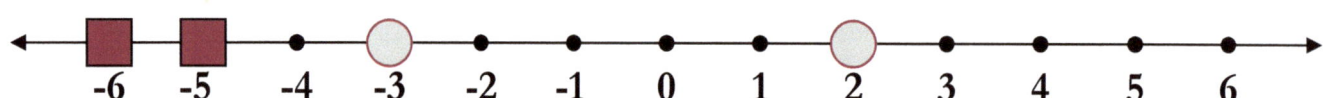

Example 1: The point (–6) lies to the left of the point (–5).
-6 < -5
-6 is less than –5

Example 2: The point (2) lies to the right of the point (–3)
2 > -3
2 is greater than –3

Practice Exercise 1: Insert < or > between each pair of numbers to make a true statement.

(a) 0   -3   (b) -5   5   (c) -4   -6   (d) 2   -5

Practice Exercise 2: Graph each pair of numbers from Practice Exercise 1 on the number line below. Color-in the circle corresponding to each number, and then write the correct number below the circle.

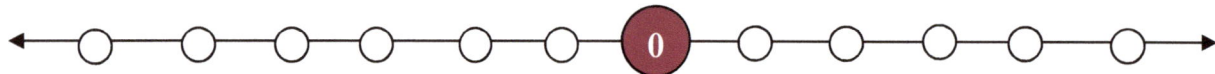

Practice Exercise 3:

a) If (5) is the number opposite of (-5) on the number line, what number is opposite of (-4)?
b) What number is opposite of (-3)?
c) What number is opposite of (2)?
d) What number is opposite of zero?

Investigation 1.2: The Number Line

Workshop 1: Signed Numbers

Math Toolbox

Example 1: Now that we have conceptualized the notion of positive and negative numbers by using a number line, we are ready to expand our signed number expertise by considering a few more applications. Earlier we used the Celsius temperature scale to explore the very cold temperatures experienced in Antarctica. The first application introduces you to two different temperature scales, one that we use most of the time in America, the Fahrenheit scale. The other temperature scale is used in science to also express very cold temperatures and is the Kelvin Scale.

There is a simple relationship between the Celsius and Kelvin Temperature scales such that converting from one to the other is very simple. If you know the temperature in Celsius, just add (+273) to get the corresponding temperature in Kelvin.

The unusual aspect of the Kelvin scale is that there are no negative temperature values. **Absolute zero** is defined by the Kelvin temperature scale as the point where all motion stops and therefore temperatures colder than absolute zero do not exist anywhere. Let's see how this works.

Figure 1.18

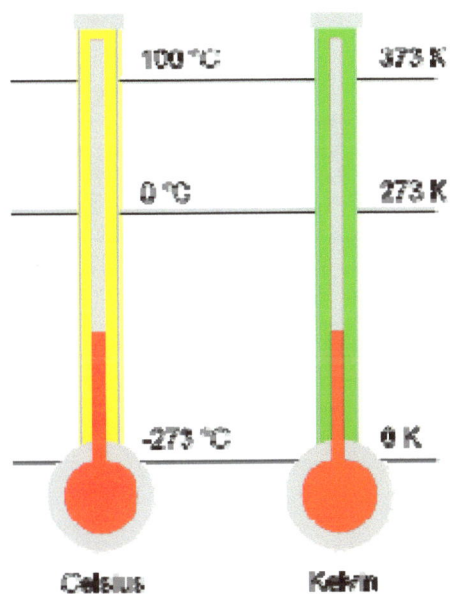

Figure 1.18 depicts side by side graphics of two thermometers. The thermometer on the left measures temperatures according to the Celsius scale and the other measures temperatures according to the Kelvin scale.

◊ ZERO Kelvin corresponds to ($-273°$) Celsius.

◊ $0°$ C corresponds to +273 Kelvin.

◊ $100°$ C corresponds to 373 Kelvin.

Because the thermometer is oriented vertically, we can think of the temperature scale depicted as a vertical line oriented up and down, rather than horizontally.

When the number line is oriented up and down the following comparison rules apply:

✱ For any two numbers graphed on a vertically oriented number line, the number above is the greater number. To indicate symbolically that one number lies above another on the number line, we use the greater than sign, >
Example: 273 K > 0 K is read, "273 Kelvin is greater than 0 Kelvin".
+273 lies **above** the number 0.

◆ The number below is the smaller number. To indicate symbolically that one number lies below another on the number line, we use the less than sign, <.
Example: $-273°$ C < $0°$ C is read, "$-273°$ C is less than $0°$ C"
-273 lies **below** the number 0

Math Toolbox

Investigation 1.2: The Number Line

Workshop 1: Signed Numbers

Water is found in three separate and distinct phases on Earth:

⛷ Ice or snow loved by skiers, but hated by commuters.

🌧 Liquid, produced as rain.

✓ Water Vapor is a gas that is a natural part of our atmosphere.

1. When the temperature of liquid water is reduced enough, it freezes, producing ice. The temperature at which ice is produced is different depending on which scale is being used. The freezing point of water on the Kelvin scale is +273 and on the Celsius scale is +0°.

2. When the temperature of liquid water is **raised** enough, water turns to vapor forming a gas. The temperature at which water vapor is produced is also different depending on which scale is being used. The boiling point of water on the Kelvin scale is +373, and on the Celsius scale is +100°.

3. There is yet a third temperature scale called the Fahrenheit scale where the boiling point and freezing point of water are different yet. On the Fahrenheit scale water boils and becomes a gas at +212° and freezes at +32°.

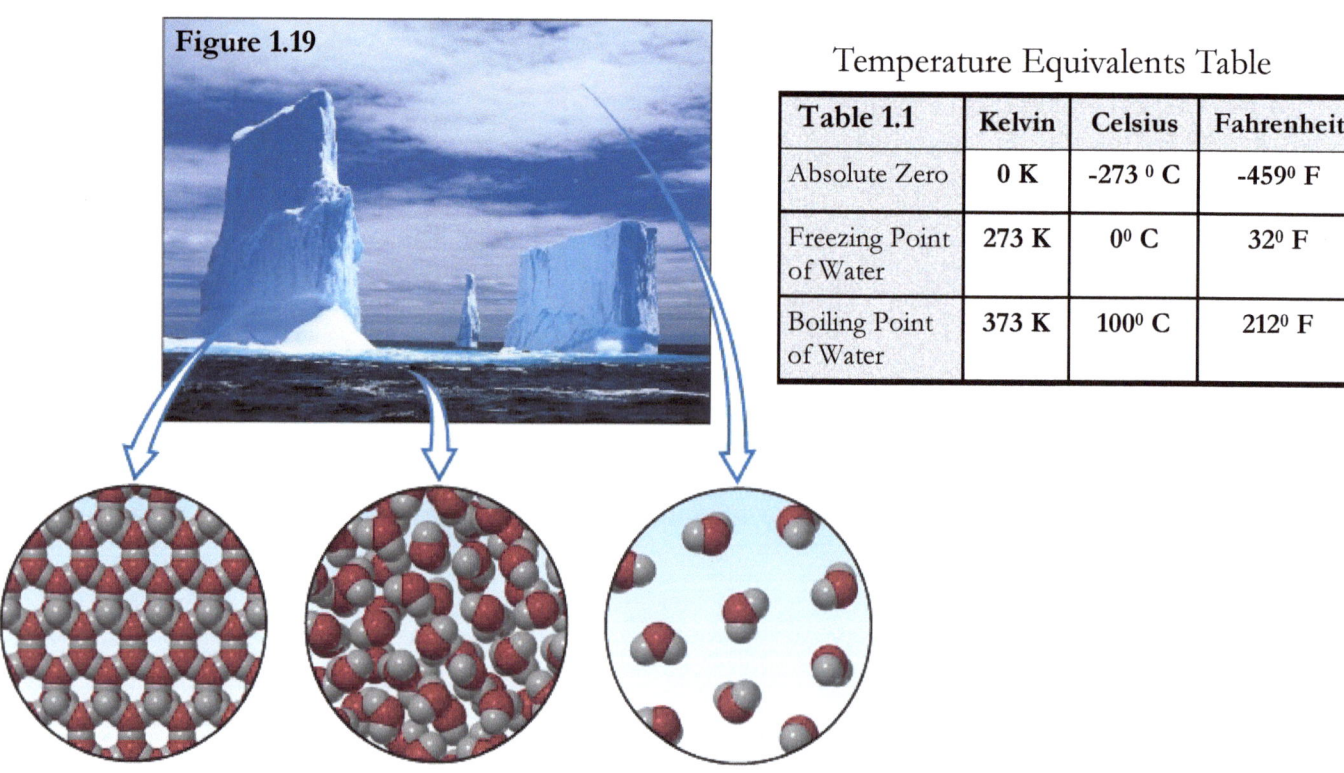

Figure 1.19

Temperature Equivalents Table

Table 1.1	Kelvin	Celsius	Fahrenheit
Absolute Zero	0 K	-273 °C	-459° F
Freezing Point of Water	273 K	0° C	32° F
Boiling Point of Water	373 K	100° C	212° F

Solid   Liquid   Gas

Water Molecules

Math Toolbox

Workshop 1: Signed Numbers

Investigation 1.2: The Number Line

Figure 1.20 shows the graph of various equivalent temperatures using the kelvin and Celsius scales.

- Notice that the values for the kelvin scale graphed along the blue line correspond to a vertical number line on the left hand side of the graph.
- The values for the Celsius scale graphed the red line correspond to a vertical line on the right hand side of the graph.

There are ten equivalent points plotted beginning with the equivalent temperatures 0 K and -273^0 C. As you move from left to right, the equivalent temperatures using the kelvin and Celsius scales increase. As an example, a pair of equivalent temperatures are circled: 150 K and -123^0 C. There are 9 other equivalent pairs shown.

Figure 1.20

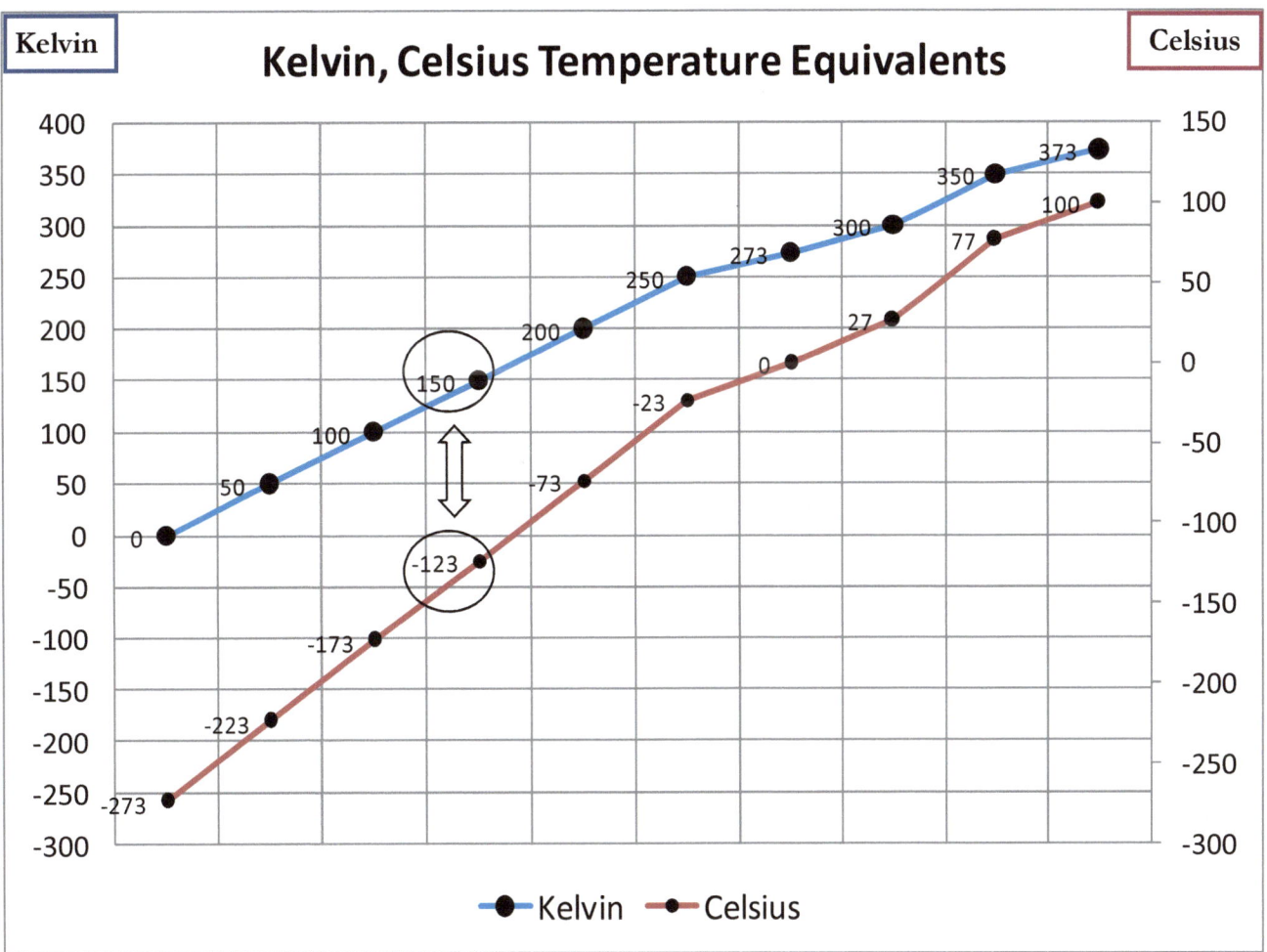

Investigation 1.2: The Number Line

Workshop 1: Signed Numbers

Figure 1.21 shows the graph of 10 equivalent temperatures using the Fahrenheit and Celsius scales.

- Notice that the values for the Celsius scale graphed along the blue line correspond to a vertical number line on the left hand side of the graph.
- The values for the Fahrenheit scale graphed the red line correspond to a vertical line on the right hand side of the graph.

There are ten equivalent points plotted beginning with the equivalent temperatures -459^0 F and -273^0 C. As you move from left to right, the equivalent temperatures using the Fahrenheit and Celsius scales increase. As an example, a pair of equivalent temperatures are circled: -223^0 C and -370^0 F. There are 9 other equivalent pairs shown on the graph.

Figure 1.21

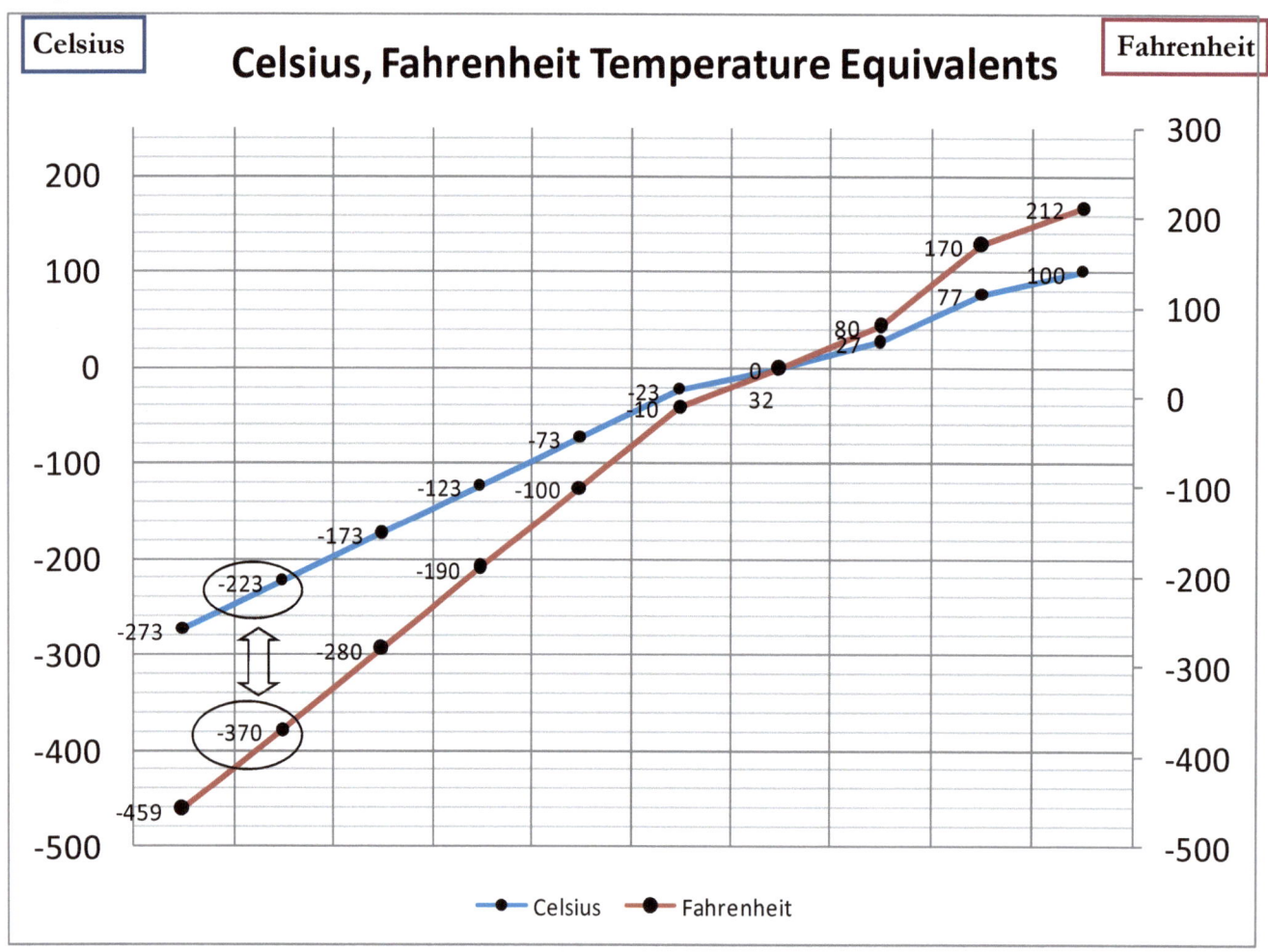

Math Toolbox

Investigation 1.2: The Number Line
Workshop 1: Signed Numbers

Temperature is measured with a thermometer calibrated to one of three temperature scales: Kelvin, Celsius, or Fahrenheit.

Intuitively, temperature is the measure of how "hot" or "cold" something is. Scientifically speaking temperature is a measure of how much energy a physical body or physical system possesses. This energy causes tiny molecules within the system or body to move. The faster the molecules move, the greater the energy, and the higher will be the temperature reading.

✓ Energy must be added to melt ice transforming it to a liquid state. As the added energy increases so does the temperature.

✓ More Energy must be added transforming the liquid to a gas. More energy = higher temperature.

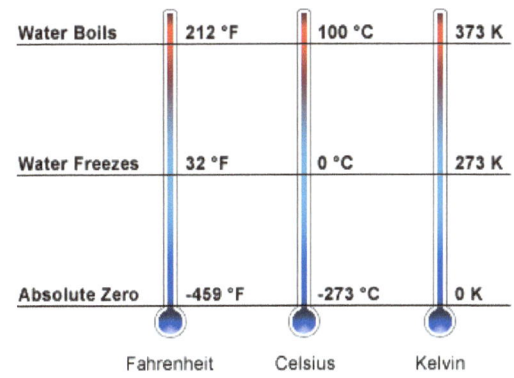

Absolute Zero   Figure 1.22

Thermometers compare Fahrenheit, Celsius, and Kelvin scales

Figure 1.23

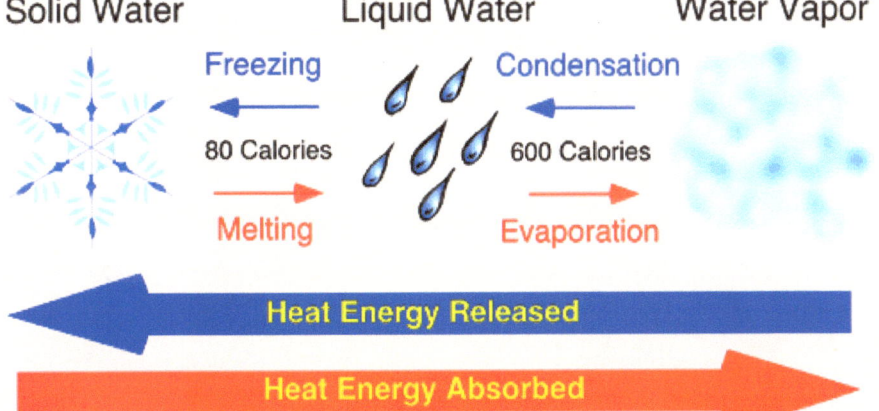

When ice is at a temperature of 0^0 C, 32^0 F, or 273 K, it freezes.

If enough energy is added to the frozen water, it eventually melts because the energy is absorbed.

When enough energy is absorbed, the water boils, turning the once frozen ice into a gas.

This process of converting water to water vapor (gas) happens when the temperature reaches 100^0 C, 212^0F, or 373 K, depending on which thermometer calibration is being used.

When water absorbs enough energy turning it into a gas, each of the three thermometer calibrations measure the same amount of energy, but use a different scale to reflect the actual temperature. Equivalent temperatures among the three thermometers mean that at 32^0 F, 0^0 C, or 273 K, the same "energy" is being measured, but the temperature is reflected based on different scales.

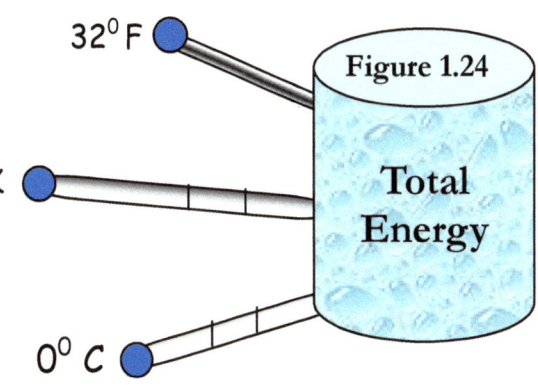

29

Math Toolbox

Investigation 1.2: The Number Line

Workshop 1: Signed Numbers

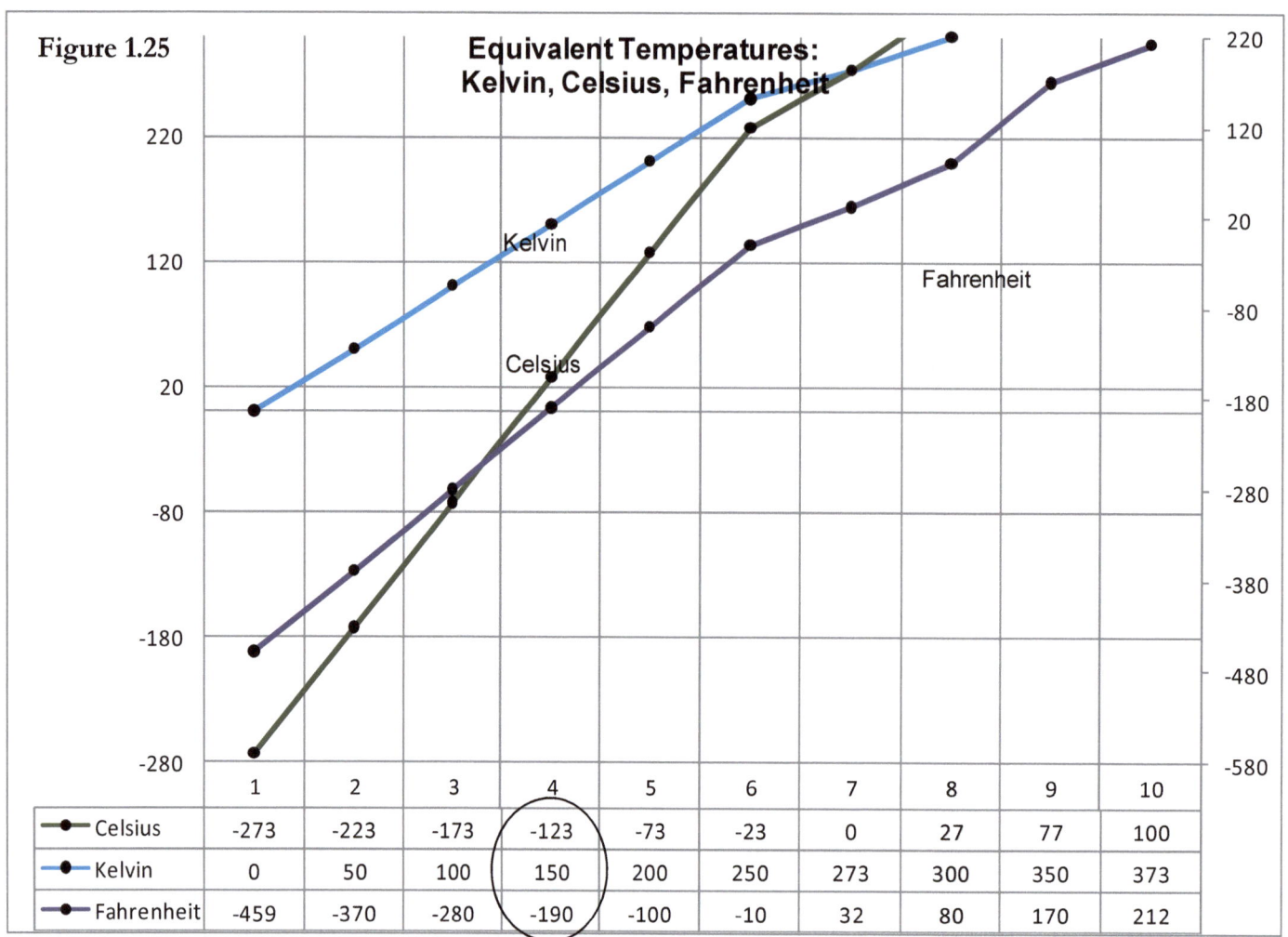

Figure 1.25

Figure 1.25 above shows three equivalent temperature lines with the corresponding data table located at the bottom.

Beginning from left to right, we see that there are 10 sets of equivalent temperatures. Each group of 3 temperatures would measure the same "heat content" but calibrated to each respective temperature scale.

Example: Set #4 contains three temperatures: $-123°$ C, 150 K, and $-190°$ F. All three of these temperatures reflect the same energy content—just calibrated to their respective temperature scales. So, in effect, all three of these temperatures have recorded the same "heat content" for the object being measured.

As you scan the bottom of the graph, there are 9 more equivalent temperature sets. As you move from left to right, from set #1 to set #10, the actual heat content increases—thus in all 10 cases the temperature increases, but according to different temperature scale calibrations.

Assignment #8 ☑ Investigation 1.2 Name

Use Figure 1.20 for Assignment #8

For this exercise, you will create a vertical number line by plotting the 10 pairs of equivalent temperatures shown on Figure 1.20 using the Celsius and Kelvin scale. A skeleton number line is drawn below. Enter the correct temperatures on each line starting with the lowest temperature value for both the kelvin and Celsius scales.

✏ For any two numbers graphed on a vertically oriented number line, the number **above** is the greater

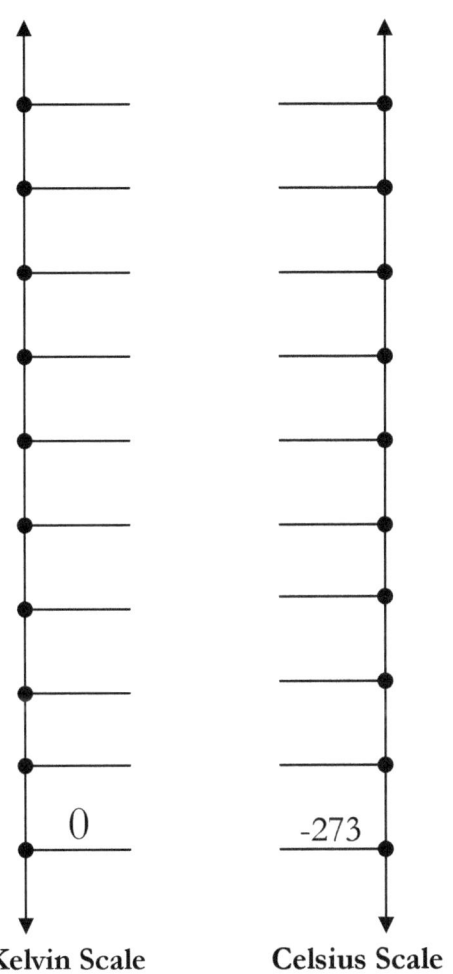

Kelvin Scale   **Celsius Scale**

Instructions: Insert < or > between each pair of temperatures to make a true statement. Pay particular attention to the symbol, ⁰C or K, included with each number.

1. –173⁰ C   -73⁰ C   2. -23⁰ C   -73⁰ C

3.   27⁰ C   -23⁰ C   4. 150 K   273 K

5. -150⁰ C   0⁰ C   6. 80⁰ C   -20⁰ C

7.   0 K   -173⁰ C   8. 0⁰ C   0 K

9. 200 K   27⁰ C   10. 77⁰ C   200 K

11. -150⁰ C   325 K   12.   0 K   100⁰ C

13. The temperature at Vostok Station in Antarctica at noon was –1⁰ C. The temperature dropped 5⁰ by 6 pm. What was the temperature at 6 pm?

14. The temperature at McMurdo Station in Antarctica was –18⁰ C at 2:00 pm. By 5:00 pm the temperature increased 3⁰. What was the temperature at 5:00 pm?

15. The temperature at the South Pole Station is –23⁰ Celsius. What is the temperature on the Kelvin thermometer?

Challenge Question 1: If the temperature on the Celsius Thermometer shows –10⁰ C and the temperature increases by 37⁰ C, what temperature does the Kelvin thermometer read?

Challenge Question 2: The temperature is dropping 2⁰ C each hour. If the starting temperature is –5⁰ C, how cold will it be in 3 hours? Is this final temperature < or > 100 K?

Assignment #9 — Investigation 1.2 Name

Use Figure 1.21 for Assignment #9

For this exercise, you will create a vertical number line by plotting the 10 pairs of equivalent temperatures shown on Figure 1.21 using the Celsius and Fahrenheit scale. A skeleton number line is drawn below. Enter the correct temperatures on each line starting with the lowest temperature value for both the Fahrenheit and Celsius scales.

👤 For any two numbers graphed on a vertically oriented number line, the number **above** is the greater number

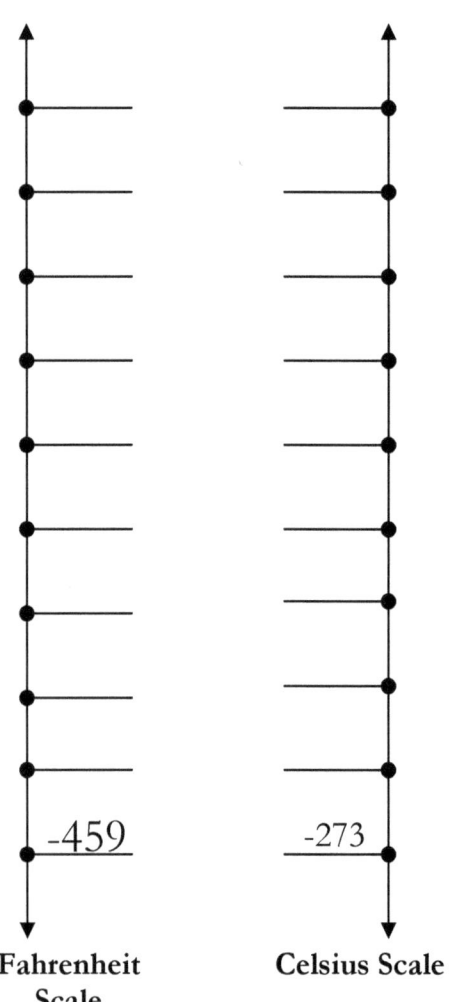

Fahrenheit Scale — −459

Celsius Scale — −273

Instructions: Insert < or > between each pair of temperatures to make a true statement. Pay particular attention to the symbol, °C or °F, included with each number.

1. −173° C   −273° C
2. −300° F   −459° F
3. −200° C   −201° C
4. −280° F   −100° F
5. −10° F   0° C
6. −10° F   −100° F
7. 0° C   −173° C
8. 0° C   60° F
9. −370° F   −123° C
10. 77° C   80° F
11. −150° C   325 K
12. 170° F   100° C

13. The temperature at Vostok Station in Antarctica at noon was −10° F. The temperature dropped 10° by 6 pm. What was the temperature at 6 pm?

14. The temperature at McMurdo Station in Antarctica was −15° F at 2:00 pm. By 5:00 pm the temperature increased 3° F. What was the temperature at 5:00 pm?

15. The temperature at the South Pole Station is −23° Celsius. What is the temperature on the Fahrenheit thermometer?

Challenge Question 1: If the temperature on the Fahrenheit Thermometer shows −20° C and the temperature decreases by 3° C, what temperature does the Fahrenheit thermometer read?

Challenge Question 2: The temperature is dropping 5° C each hour. If the starting temperature is −15° C, how cold will it be in 3 hours? Is this final temperature < or > −10° F ?

✎ Assignment #10   ✓ Investigation 1.2   Name

Use Figure 1.25 for Assignment #10

For this exercise, you will create a horizontal number line by plotting the 10 sets of equivalent temperatures shown on Figure 1.25 using the Kelvin, Celsius, and Fahrenheit scale. Three separate skeleton number lines are drawn below. Enter the correct temperatures on each of the three lines starting with the lowest temperature value for all three scales. Then use the equivalent values provided in the table at the bottom of the graph of Figure 1.25 to complete your number lines.

🗣 For any two numbers graphed on a horizontally oriented number line, the number **to the right** is the greater number.

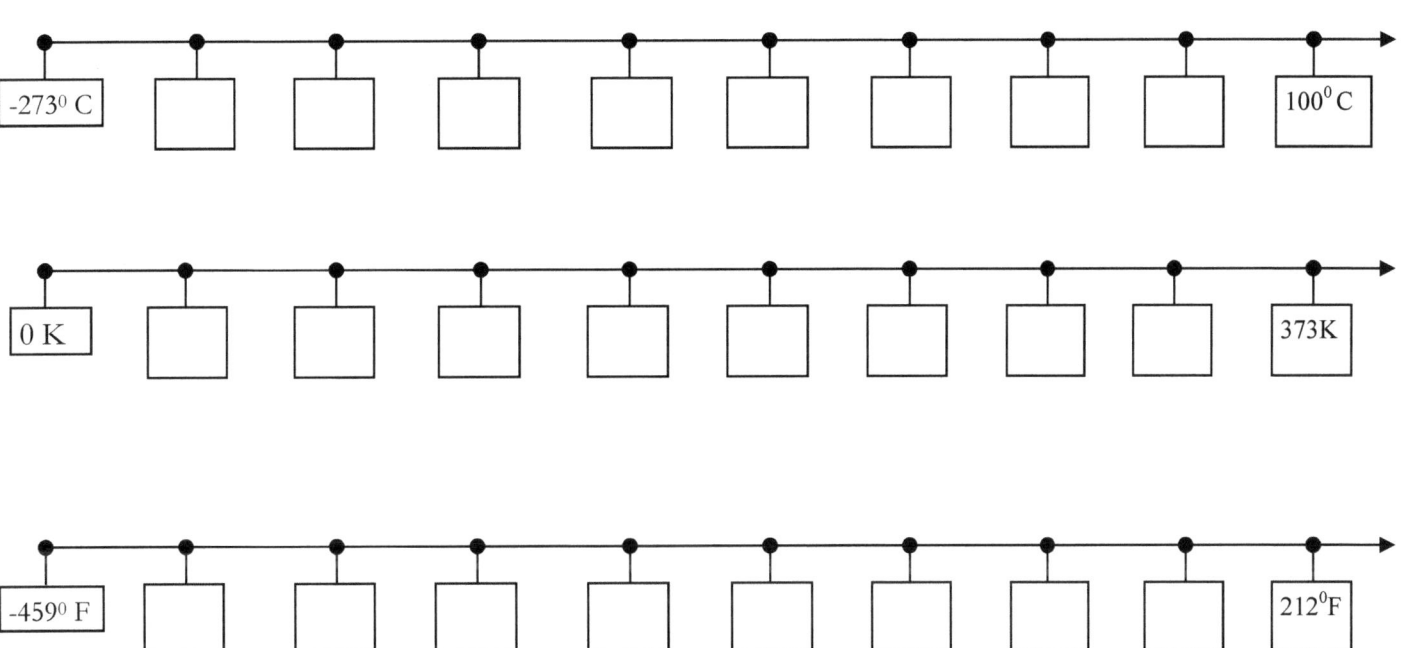

Instructions: Use the completed number lines above and insert the correct sign, <, >, or =, to make each of the following statements true.

1. 0 K   -370° F   2. 100°C   212° F   3. -459° F   -450° F   4. -23° C   273 K

In problems 5 through 10, you are asked to insert two correct signs. Example: 100 K < 32° F < 27° C
This is read: 100 K **is less than** 32° F, and 32° F **is less than** 27° C.

5. -273°C   0 K   -190°F   6. 212° F   100° C   373 K   7. 170° F   300 K   0° C

8. 50 K   -173° C   -100° F   9. -10° F   -50° F   -75° F   10. 0 K   20 K   10 K

Challenge Question #1: If the starting temperature is –73° C and the temperature increases by 50° C, what temperature is read on the Celsius, Kelvin and Fahrenheit thermometers?

✎ Assignment #11   ☑ Investigation 1.2   Name

Instructions: All of the statements below are false. Cross out the incorrect word, numbers or symbol. Then enter the correct word, numbers or symbol in the box to make each a true statement.

1. The Dow Jones stock market average fell 50 points on Monday. On Tuesday it fell an additional 75 points. Altogether the Dow Jones stock market average fell ~~25~~ points on Monday and Tuesday.   **1.** 125

2. At the McMurdo Station Antarctica, the temperature at midday was $-15°$ Celsius. By 5:00 pm the temperature was $-35°$ degrees. In five hours time the temperature fell ~~25°~~.   **2.** 20°

3. A miner begins his day underground at -100 feet. During the day the miner is transported an additional -125 feet underground. When his shift is over he must travel ~~25~~ feet up to ground level.   **3.** 225

4. A SCUBA diver descends to 40 feet below the ocean's surface. While resting at this distance below sea level, she sees a companion diver 20 feet below her. To join him, her new depth underneath the water will be ~~-70~~ feet below sea level.   **4.** -60

5. ~~$-3°$ C~~ is always warmer than $5°$ Celsius.   **5.** colder

6. There are ~~many~~ temperatures below absolute zero—0 K.   **6.** no

7. From the point $+3$ on a number line, one would move 5 units to the left to end up at the number ~~$+8$~~. (Hint: draw a number line).   **7.** -2

8. From the point $+10$ on a number line, one would move 5 units to the left to end up at ~~-15~~. (Hint: Draw a number line)   **8.** $+5$

9. A SCUBA diver is swimming 35 feet below the surface of the water in the Gulf of Mexico. He must swim ~~-35~~ feet upward to reach the surface.   **9.** $+35$

10. 0 Kelvin and ~~$-459°$ C~~ are equivalent temperatures.   **10.** $-273°$ C

11. On a horizontal number line beginning at the point $-10°$C, one must move ~~10~~ units to the right to reach $20°$ C.   **11.** 30

12. On a vertical number line beginning at the point $-30°$ F, one must move ~~down~~ 30 units to reach $0°$ F.   **12.** up

13. The freezing point of water is ~~0 K~~ and $0°$ Celsius.   **13.** 273 K

14. The boiling point of water is $212°$ F, $100°$ C, and ~~273~~ K.   **14.** 373

☑ Skill Builder #3   Name

Choose all numbers from each given list that make each statement true.

1. ☐ < -9   A. 0   B. –6   C. -10   D. 9
                  E. –12   F. 10   G. –20   E. 1

2. -4 > ☐   A. 0   B. –6   C. –8   D. 9
                  E. –12   F. –5   G. –20   E. 1

Answer true or false for each statement. If your answer is false, give one example that supports that conclusion.

In each of the problems below you are given a list of numbers. Draw a number line and graph each integer in the list on the same number line.

3. If a > b, then "a" must be a positive number.

10. 1, 2, 4, 6

4. A Positive number is always greater than a negative number.

11. 1, -1, -2, -4, 3, 5

5. A negative number never lies to the left of a positive number on the number line.

12. 0, -7, 3, -6

6. The sum of 9 + 3 lies to the right of –12 on the number line.

13. 0, 3, 6, -4, 2, -2

7. If the number "a" lies to the left of the number "b" on a number line, then "a" must be a negative number.

14. -5, 3, 0, 8, -6

8. The freezing point of water on the Kelvin scale is +273 and on the Celsius scale is +0°.

15. 0, -2, -7, 5, -10

9. 27° C and 80° F are equivalent temperatures.

Math Toolbox
Workshop 1: Signed Numbers
Investigation 1.3: Adding Integers Using a Number Line

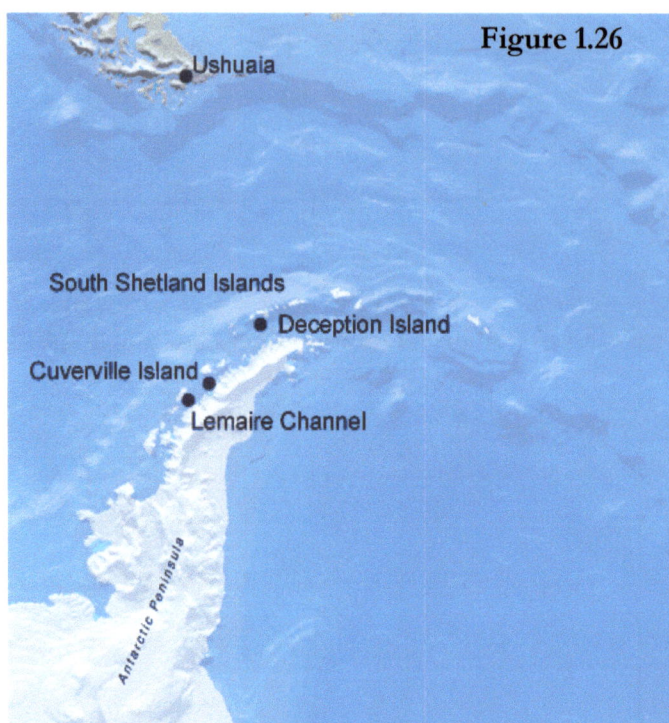

Figure 1.26

Figure 1.26 depicts the Antarctic peninsula and Antarctic islands, some of the last truly unspoiled regions of the world.

Known as the Mysterious White Continent, enormous numbers of penguins, whales, seals, and seabirds congregate in the waters off the Antarctic Peninsula and associated islands to hunt for a plentiful food supply. In addition to its unspoiled ecological beauty, the Antarctic Peninsula is adorned with multi–colored ice caps, glistening glaciers, and snow-capped mountains towering overhead.

But along with its stunning beauty, the cold still persists. The average temperatures for the Antarctic Peninsula from January through December is shown in Table 1.2 below, where the averages are rounded to the **nearest integer**.

Table 1.2 Antarctic Peninsula Average Temperature Rounded to the Nearest Celsius Integer												
Month	Jan	Feb	Mar	Apr	May	Jun	Jul	Aug	Sep	Oct	Nov	Dec
Temperature	1	0	-2	-4	-7	-9	-13	-12	-9	-7	-3	1

What's an Integer?

Integers are a collection (or set) of numbers that contain both positive and negative numbers, without decimals or fractions. The complete set of integers is written using *set notation* that consists of braces enclosing the largest and smallest values. **The <u>complete set</u> of integers using set notation can be written:**

$$\{-\infty \ldots -7, -6, -5, -4, -3, -2, -1, 0, 1, 2, 3, 4, 5, 6, 7, \ldots +\infty\}$$

∞ The three dots on both ends of the integer list mean that the numbers continue to increase or continue to decrease to infinity; there is no "largest" number, or "smallest number".

But a set doesn't necessarily have to contain all possible numbers in the integer set shown above. For instance, we can create a set of integer temperatures from Table 1.2 above. Notice that the integer 1 is repeated as an average temperature for both December and January. We will only list it one time. -9 is also repeated twice. Below is the set of average temperature integers for the Antarctic Peninsula written from smallest to largest.
There are nine members or elements of the average temperature set.
$$\{-13, -12, -9, -7, -4, -3, -2, 0, 1\}.$$

Investigation 1.3: Adding Integers Using a Number Line

Workshop 1: Signed Numbers

Sometimes it is beneficial when adding integers together to utilize a number line.

Example 1: Consider the following scenario. The temperature on the ground at Vostok Research Station is -3^0 C and the temperature drops -5^0 C. To get the final temperature, we must add these two numbers together. Add $(-3) + (-5)$ using a number line.

Step 1: Begin at (-3).

Step 2: Move 5 units to the left to simulate adding the number, -5.

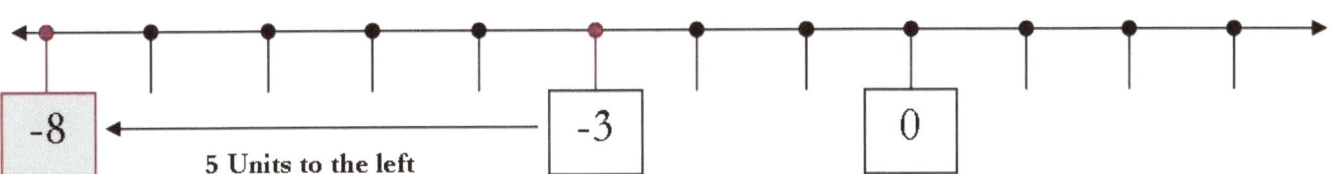

Step 3: Count the number of places from zero. Since we are to the "left" of zero, the final answer will be negative.

Step 4: Write your answer: $(-3) + (-5) = (-8)$

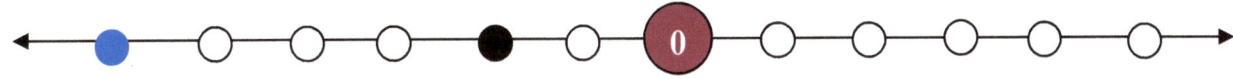

Example 2: Using the number line above, add these integers: $(-2) + (-4) + (+6)$

Step 1: Begin at (-2) ●

Step 2: Move 4 units to the left to simulate adding the number, (-4). ●

Step 3: Move 6 units to the **right** since we are adding the number (+6). (End up at zero). ⓪

Step 4: Write your answer: $(-2) + (-4) + (+6) = 0$

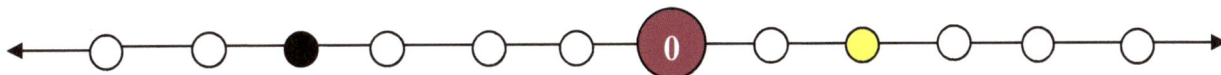

Example 3: Add the same numbers, but change their order. Add: $(-4) + (+6) + (-2)$

Step 1: Begin at (-4) ●

Step 2: Move 6 units to the right since we are adding the number (+6). ○

Step 3: Move 2 units to the left to simulate adding the number (-2). (End up at zero). ⓪

Investigation 1.3: Adding Integers Using a Number Line

Workshop 1: Signed Numbers

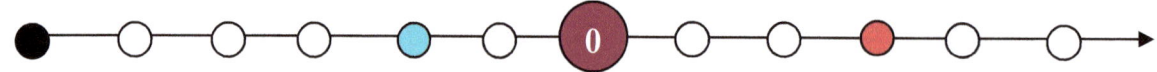

Example 4: On January 6 the temperature at Caribou, Maine at 8 a.m. was -6^0 F. By 9 a.m. the same morning the temperature had risen 4^0 F. By 10 a.m. the temperature rose another 5^0 F from the 9 a.m. temperature.

⇒ **What was the temperature at 10 a.m.?** Find the temperature by using the number line.

Step 1: Begin at –6 ●

Step 2: Move 4 units to the right, since the temperature had "risen" 4 units. ○ This is the temperature at 9 a.m.

Step 3: Move 5 units to the right, since the temperature had "risen" an additional 5 units. ○

Step 4: Count the number of units from zero. Because the ending position is "3" units to the right of zero, our answer is $(+3^0$ F).

Step 5: Write the equation: $(-6) + (+4) + (+5) = (+3)$.

Step 6: Write your answer as a sentence. The temperature at 10 a.m. was $+3^0$ F.

Practice Exercises: Draw your own number line, labeling each of 10 units to the right and to the left of zero. Use this number line to add the following integers. Write your answer in the box provided.

1. $(-5) + (-2) + 11 = $ ☐

2. $(+5) + (+2) + (-11) = $ ☐

3. $(-6) + (+3) + (-1) + (-2) = $ ☐

4. $(+6) + (-3) + (+1) + (-5) = $ ☐

5. $(-2) + (-3) + (+4) + (-5) + (6) = $ ☐

Math Toolbox

Workshop 1: Signed Numbers

Investigation 1.3: Adding Integers Using a Number Line

Table 1.2 Antarctic Peninsula Average Temperature Rounded to the Nearest Celsius Integer												
Month	Jan	Feb	Mar	Apr	May	Jun	Jul	Aug	Sep	Oct	Nov	Dec
Temperature	1	0	-2	-4	-7	-9	-13	-12	-9	-7	-3	1

Example 5: Using Table 1.2, the Antarctic Peninsula Average Temperature Integers for the year, draw a number line and verify the answers to the following questions.

1. Did the average temperature increase or decrease from January to February? By how much? Write a true statement using < or > for the given temperatures.

    ⇒ The average temperature decreased 1^0 C.
    ⇒ 1^0 C > 0^0 C

2. Did the average temperature increase or decrease from March to July? By how much?

    ⇒ The average temperature in March was -2^0 C. In July it was -13^0 Celsius. On a number line we would move 11 units to the left when going from (–2) to (–13). Therefore, the average temperature decreased by 11^0 C.
    ⇒ -2^0 C > -13^0 C

3. Did the average temperature increase or decrease from April to October? By how much?

    ⇒ The average temperature in April was -4^0 C. In October the average temperature was -7^0 C. Beginning at –4 on a number line, we would move 3 units to the left when going from (-4) to (-7). Therefore, the average temperature decreased by 3^0 C.
    ⇒ -4^0 C > -7^0 C

4. Did the average temperature increase or decrease from September to March?

    ⇒ The average temperature in September was -9^0 C. In March the average temperature was -2^0 C. Beginning at -9^0 C on a number line, we would move 7 units to the right when going from (–9) to (–2). Therefore, the average temperature increased by 7^0 C.
    ⇒ -9^0 C < -2^0 C

✎ Assignment #12 ☑ **Investigation 1.3** Name
Page 1 of 2

In golf, **par** is an average, predetermined number of strokes that a golfer should need to complete a hole. Scores at the end of play are either *over par* which are interpreted as "positive values" or *under par* interpreted as "negative values". To be over par means that a golfer required more strokes than are considered average for to make a hole (this is a bad thing). To be under par means that a golfer required less strokes than are considered average for a given hole (this is a good thing).

Example: Assume a particular hole would, on average, require 4 strokes (# times the golfer swings the club) to go from the golf tee to the hole. If the golfer actually had to swing the club 5 times in order to get the ball in the hole, the score for the golfer would be 1 "over par". This would be recorded as +1. If another golfer only required 3 strokes to reach the hole, the score would be recorded as –1, read 1 "under par". The lower the golfer's score, the better.

Mary and Matilda are contending for the best score on 9 holes of golf. For all 9 holes on this particular golf course, Par is 36. The ladies have decided that after 9 holes of golf, the one with the lowest score will buy lunch. The table below indicates the following: (i) Par for each hole (ii) Mary's score for each hole and (iii) Matilda's score for each hole.

Figure 1.27

"We've only played three holes and we've both already shot our ages!"

✎ Before answering the questions on the next page, use a number line to add integers and enter your answers in the empty cells in the table below.

✎ Using a number line, add the integers and complete the "Score" column in the table below by (i) adding up Mary's score, (ii) adding up Mary's total number of strokes (iii) adding up Matlida's score and (iv) adding up Matilda's total number of strokes.

Table 1.3	Hole 1	Hole 2	Hole 3	Hole 4	Hole 5	Hole 6	Hole 7	Hole 8	Hole 9	Score
Par	4	3	5	5	4	3	4	4	4	
Mary	0	+1	+2	0	–1	+1	–1	0	+1	
Total Strokes										
Matilda	0	–1	+1	–2	0	+1	0	+1	–1	
Total Strokes										

40

✎ Assignment #12
Page 2 of 2   ☑ Investigation 1.3   Name

1. Draw a number line and use it to total Mary's score from Table 1.3. Write your answer in the box provided.

2. Using your number line, find a total for Matilda's score using Table 1.3. Write your answer in the box provided.

3. Using Mary's score from Table 1.3 and knowing that "par" for this golf course is 36 strokes, how many strokes did Mary need to complete the 9 holes of golf?

4. Using Matilda's score from Table 1.3 and knowing that "par" for this golf course is 36 strokes, how many strokes did Matilda need to complete 9 holes of golf?

5. Who bought lunch?

6. Which golfer, Mary or Matlida, had a better score on hole #1? Write their scores on hole #1 using <, >, or =.

7. Which golfer, Mary or Matlida, had a better score on hole #2? Write their scores on hole #2 using <, >, or =.

8. Which golfer, Mary or Matlida had a better score on hole #3? Write their scores on hole #3 using <, >, or =.

9. Which golfer, Mary or Matlida had a better score on hole #4? Write their scores on hole #4 using <, >, or =.

10. Which golfer, Mary or Matlida had a better score on hole #5? Write their scores on hole #5 using <, >, or =.

11. Which golfer, Mary or Matlida had a better score on hole #6? Write their scores on hole #6 using <, >, or =.

12. Which golfer, Mary or Matlida had a better score on hole #7? Write their scores on hole #7 using <, >, or =.

13. Which golfer, Mary or Matlida had a better score on hole #8? Write their scores on hole #8 using <, >, or =.

14. Which golfer, Mary or Matlida had a better score on hole #9? Write their scores on hole #9 using <, >, or =.

✎ Assignment #13
Page 1 of 4

☑ **Investigation 1.3**

Name

Term	Integer Score	Definition	Table 1.4
Bogey	+1	1 stroke "over" par	
Birdie	-1	1 stroke "under par"	
Eagle	-2	2 strokes "under par"	
Double Bogey	+2	2 strokes "over par"	
Triple Bogey	+3	3 strokes "over par"	

A bit of history... In the early part of the 20th Century, the word *par* began to be applied to the ideal score of professional golfers. The term **bogey** gradually began to describe the ideal score of recreational golfers. From there, it was a short leap to its current definition for a score of 1-over par. As "par" became the accepted term for *a good* score on a hole, "bogey" was applied to the higher score recreational golfers might expect to achieve, that is the *worse* score. Table 1.4 above lists terms that might be used to describe a particular number of strokes in a golf game.

On a standard 9 hole, par 36 golf course, Bob and Ben are also playing to see who buys lunch. They, like the ladies in Assignment #12, have agreed that the golfer with lower score at the end of play on 9 holes of golf will buy lunch. Figure 1.28 and Figure 1.29 on the following page tracks each player's integer score from hole #1 to hole #9 and records it as an integer score over or under par. Use the information on the graph to complete the table below by inserting the appropriate integer score and correct term describing each score.

Table 1.5

Player	Hole 1	Hole 2	Hole 3	Hole 4	Hole 5	Hole 6	Hole 7	Hole 8	Hole 9
Bob									
Term (bogey, etc)									
Ben									
Term (bogey, etc)									

Assignment #13
Page 2 of 4
☑ **Investigation 1.3**
Name

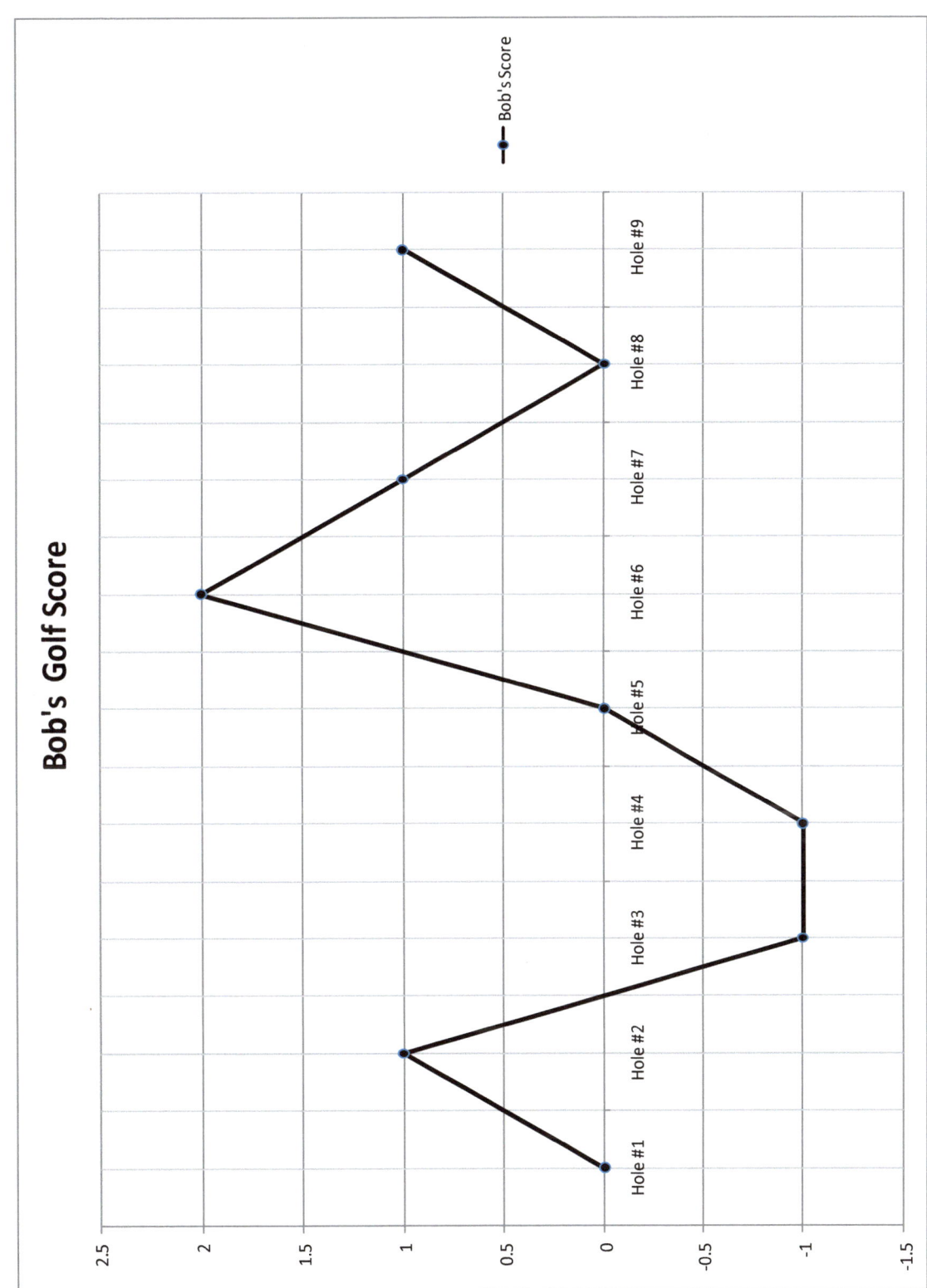

Figure 1.28

✎ Assignment #13

☑ **Investigation 1.3**   Name

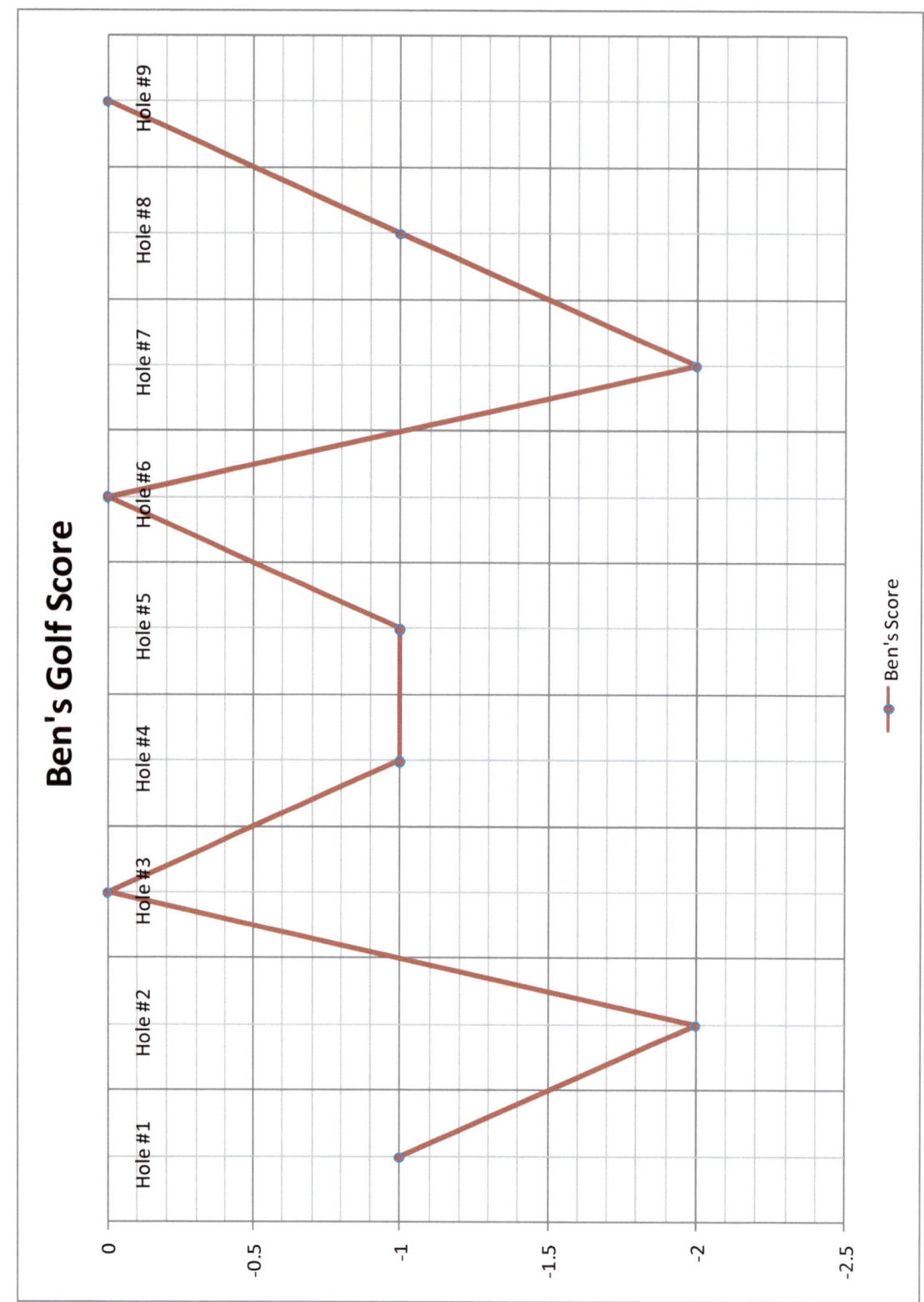

Figure 1.29

✎ Assignment #13
Page 4 of 4

☑ **Investigation 1.3**

Name

Instructions: Adding both down and across, **fill in all empty cells**. You will need the results of Table 1.5.

✎ After determining the total number of strokes, per hole, for Bob and Ben separately, calculate by adding down the total number of strokes for Bob and Ben together.

✎ Calculate the total number of strokes Bob required to finish all 9 holes. (Add across)

✎ Calculate the total number of strokes Ben required to finish all 9 holes. (Add across).

Table 1.6	Hole 1	Hole 2	Hole 3	Hole 4	Hole 5	Hole 6	Hole 7	Hole 8	Hole 9	Total
Par	4	4	3	5	3	4	5	4	4	36
Bob's total # Strokes										
Ben's Total # Strokes										
Total										

Add Down →

Add Across →

1. Write the relationship between the total # of strokes for Ben and Bob using <, >, or =.

2. Is it clear from your answer to question #1 which player is the professional? Why?

3. Who bought lunch?

4. How many strokes did Bob and Ben make together, to complete hole #2?

5. How many strokes did Bob and Ben make together, to complete hole #5?

6. Write the total scores Bob and Ben make together as an expression using <, >, or =.

☑ Skill Builder #4   Name

Add the following integers. Then graph your result on the number line provided.

1. 8 + (-2)   ←—————————————————→

2. -13 + 7   ←—————————————————→

3. 10 + (-4)   ←—————————————————→

4. (-6) + (-5)   ←—————————————————→

5. -6 + 0   ←—————————————————→

6. (-4) + (-2)   ←—————————————————→

7. (-63) + (63)   ←—————————————————→

8. 15 + 42   ←—————————————————→

9. 5 + (-9)   ←—————————————————→

10. (-123) + (-100) ←—————————————————→

11. (-25) + 30   ←—————————————————→

12. (-10) + 9   ←—————————————————→

☑ Skill Builder #5   Name

Add the following numbers.

1. -4 + 2 + (-5)

2. -1 + 5 + (-6)

3. 12 + (-4) + (-4) + 4

4. (-10) + 14 + 6 + (-10)

5. 3 + (-23) + 19

6. -10 + (-6) + (-1)

7. (-45) + 20 + 23

8. -37 + 47 + (-10) + (-5)

9. -16 + 6 + (-14) + (-10)

10. 125 + (-100) + (-50) + 25

11. -80 + (-10) + 6 + 20

21. 13 + (-20) + 6 + (-5)

22. 18 + (-9) + 5 + (-3)

23. 6 + (-17) + 1

24. 23 + (-3) + 20 + (-40)

25. 4 + 3 + (-15)

26. -11 + 5 + (-16)

27. 2 + (-14) + (-2) + (-1)

28. 13 + (-16) + 6 + (-8)

29. 13 + (-20) + 9

30. -10 + (6) + (-1) + 5

31. (-35) + 10 + (-14)

32. 3 + 47 + (-10) + (-50)

33. -16 + 16 + (-14) + 18

34. -125 + (-100) + (-150) + 225

35. -60 + (-10) + 50 + 25

36. 15 + (-20) + 16 + (-5)

37. 18 + (-9) + 25 + (-30)

38. 16 + (-17) + 32

39. 23 + (-3) + 20 + (-40)

☑ **Skill Builder #6**   Name _____

Solve the following Word Problems.

1. The temperature at noon was -8^0 C. Two hours later at 2 pm the temperature had risen 4^0 C. What is the temperature at 2 pm?

2. The temperature at noon was $+8^0$ C. Four hours later at 4 pm the temperature had dropped 14^0 C. What is the temperature at 4 pm?

3. Bob and Ben played a series of three golf games. Their scores are shown on the table at the right. Determine the total score above or below Par for both Bob and Ben. Who had the "best" score in Golf?

	Game 1	Game 2	Game 3	Game 4
Bob	+4	-2	+1	-3
Ben	-3	+2	0	-2

3. The table at the right shows the scores for two teams playing a series of card games, where it is possible to have positive and negative scores. Use the table to answer the following questions:

	Game 1	Game 2	Game 3	Game 4
Team 1	-2	-20	+19	4
Team 2	15	-11	-7	-3

(a) Find each team's total score after four games. If the winning team is the team with the greater score, find the winning team.

(b) Find each team's total score after three games. If the winner is the team with the *lowest score*, which team was winning after three games?

(c) At the halfway point (after game #2) how many total points did each team have? If the winner is the team with the greater score which team is ahead. How far ahead is this team?

(d) Consider Game #2 and Game #3 for this question. Assume that the winner has the lowest score. After playing Game #2 and Game #3, which team as the lowest score?

Skill Builder #7   Name

Team 1	Hole 1	Hole 2	Hole 3	Hole 4	Hole 5	Hole 6	Hole 7	Hole 8	Hole 9	Score
Par	4	3	5	5	4	3	4	4	4	
Mary	0	-1	-2	+1	-2	+2	-5	-1	0	
Total Strokes										
Matilda	+1	-2	11	+2	-1	0	-2	0	-1	
Total Strokes										

Team 2	Hole 1	Hole 2	Hole 3	Hole 4	Hole 5	Hole 6	Hole 7	Hole 8	Hole 9	Score
Par	4	3	5	5	4	3	4	4	4	
Debbie	+1	-2	0	-1	-3	+2	0	-1	+1	
Total Strokes										
Dafney	+2	0	3	-2	-1	-3	+2	0	+2	
Total Strokes										

Mary and Matilda are co-members of team #1. Debbie and Dafney are co-members of team #2. Each team plays 9 holes on the same day. Determine the number of strokes for each player, on each hole, and total their score. Use the numbers from these tables to answer the questions below.

1. If the winners are the team with the lower score, which team was winning after hole #5?

2. Add the total strokes for Team #1. Add the total strokes for Team #2. Write these two numbers as an inequality using < or >.

3. Add the total Score for each team (the highlighted yellow and green values). Write these two numbers as an inequality using < or >.

4. Which team was "winning" after hole #7? By how much?

5. Which team was "winning" after hole #3? By how much?

Math Toolbox

Workshop 1: Signed Numbers

Investigation 1.4: Adding Signed Integers

The real world, it turns out, is a wonderful laboratory for learning to use signed numbers. And no laboratory is better understood by the adult learner than the one represented by money. To capitalize on this knowledge that unites nearly everyone old enough to write a check, Investigation 1.4 will take you through a series of examples that illustrate both the motivation and the rules for adding and subtracting signed numbers.

Let's Begin by understanding what is inferred by a positive value versus a negative value as they are attached to the concept of money.

Example 1: Wanda opens a checking account with a starting balance of $200. On a single day Wanda makes two withdrawals of $40 and $10 respectively. Because these withdrawals indicate that money has been subtracted from Wanda's balance, they both come with a negative sign.

Withdrawal #1: -$40   **Withdrawal #2: -$10**

- ✓ Each withdrawal is a debit and is preceded by a negative sign.
- ✓ Each deposit is a credit and is preceded with a plus sign.

Each of the two withdrawals from Wanda's checking account are debits. Let's add the two "debits" together:

(-$40) + (-$10) = -$50.

| The Sum of two or more *negative numbers* will always result in a negative number. |

After adding the two debits to Wanda's checking account, we find the following balance:

+$200 + (-$40) + (-$10) = $200 + (-$50) = $150

✓ **Knowledge Check 1:**

Add the following "debits" together:

(a) (-$40) + (-$30) = _____
(b) (-$10) + (-$5) + (-$15) = _____
(c) (-$1) + (-$4) + (-$10) + (-$3) = _____

The solutions to the above knowledge check exercises should convince you that when two or more negative numbers are added together, the result will *always* be a negative number.

Solution: (a) (-$70)   (b) (-$30)   (c) (-$18)

We now consider the opposite scenario, adding two or more positive numbers together. In our checking account example, positive numbers are associated with credits, thus increasing the amount of money Wanda has on hand in her account.

Wanda deposits two "credits" to her account::

Deposit #1: +$100   **Deposit #2: +$40**

Each of the two credits are added together:

(+$100) + (+$40) = +$140

When adding two or more positive numbers together, the second truism emerges:

| The Sum of two or more *positive numbers* will always result in a positive number. |

50

Math Toolbox

Workshop 1: Signed Numbers

Investigation 1.4: Adding Signed Integers

✓ **Knowledge Check 2:**

Add the following "credits" together:

(a) (+$40) + (+$30) = _____
(b) (+$10) + (+$5) + (+$15) = _____
(c) (+$1) + (+$4) + (+$10) + (+$3) = _____

The solutions to the above knowledge check exercises should convince you that when two or more positive numbers are added together, the result will *always* be a positive number.

Solution: (a) (+$70)   (b) (+$30)   (c) (+$18)

Perhaps by now you are asking the all important question, "But what happens when I mix the two together, adding both positive and negative numbers?"

From previous work done in Workshop 1, you have seen that there is no ready formula for determining the sign of the sum. That's why you were introduced to the number line as a way to logically add signed numbers together.

When "adding" both positive and negative integers, together, the answer depends on the relative distance each of the numbers is from zero.

It's much like the tug-of-war illustrated in the graphic above. The positive integers are on one side, the negative integers are on the other. In one scenario, both sides may cancel each other out. The sum will be zero. If the negative and positive values are both the same distance from zero, then the sum will be zero.

Example 2: Wanda has a balance in her checking account of $290 after all the debits and credits have been added.

That places the $290 on the right hand side of the number line. From here, if Wanda withdraws $290 from her checking account, (adding a debit of -$290), she will move 290 places to the left on the number line, bringing her balance back to zero.

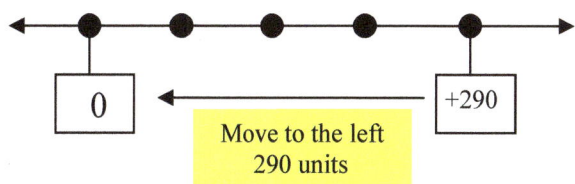

Move to the left 290 units

$$(+\$290) + (-\$290) = 0$$

✳ These two numbers, one positive and one negative, are called additive inverses of each other because they lie the same distance from zero on opposites sides of the number line.

✳ Anytime additive inverses are added together, the result is zero.

✓ **Knowledge Check 3:**

Add the following integers. Each integer is an additive inverse of the other:

(a) (-3) + (+3) = _____
(b) (-100) + (+ 100) = _____
(c) (-55) + (+55) = _____
(d) (-25) + (+25) = _____

When two integers are the same distance from zero on the number line, but lie on opposite sides of it are additive inverses of each other.

When additive inverses are added together, their sum is zero.

51

Math Toolbox

Workshop 1: Signed Numbers

Investigation 1.4: Adding Signed Integers

Example 2: *Wanda's Widgets*—Month #1

Figure 1.30, a column chart for Month #1, tracks the debit and credit activity during Month #1 for Wanda's new business checking account. She has gone into business producing widgets and will be collecting revenue from the sales deposited as "credits" to her account. But with every business there are also inevitably the costs incurred to actually produce Wanda's widgets. These costs will also be paid from the business checking account and are recorded in the graph below as "debits".

[1] The **height** of each column provides a visual clue into Wanda's early success during Month #1.

🗣 Does she appear to have more "credits" during Month #1 or more "Debits"?

Figure 1.30

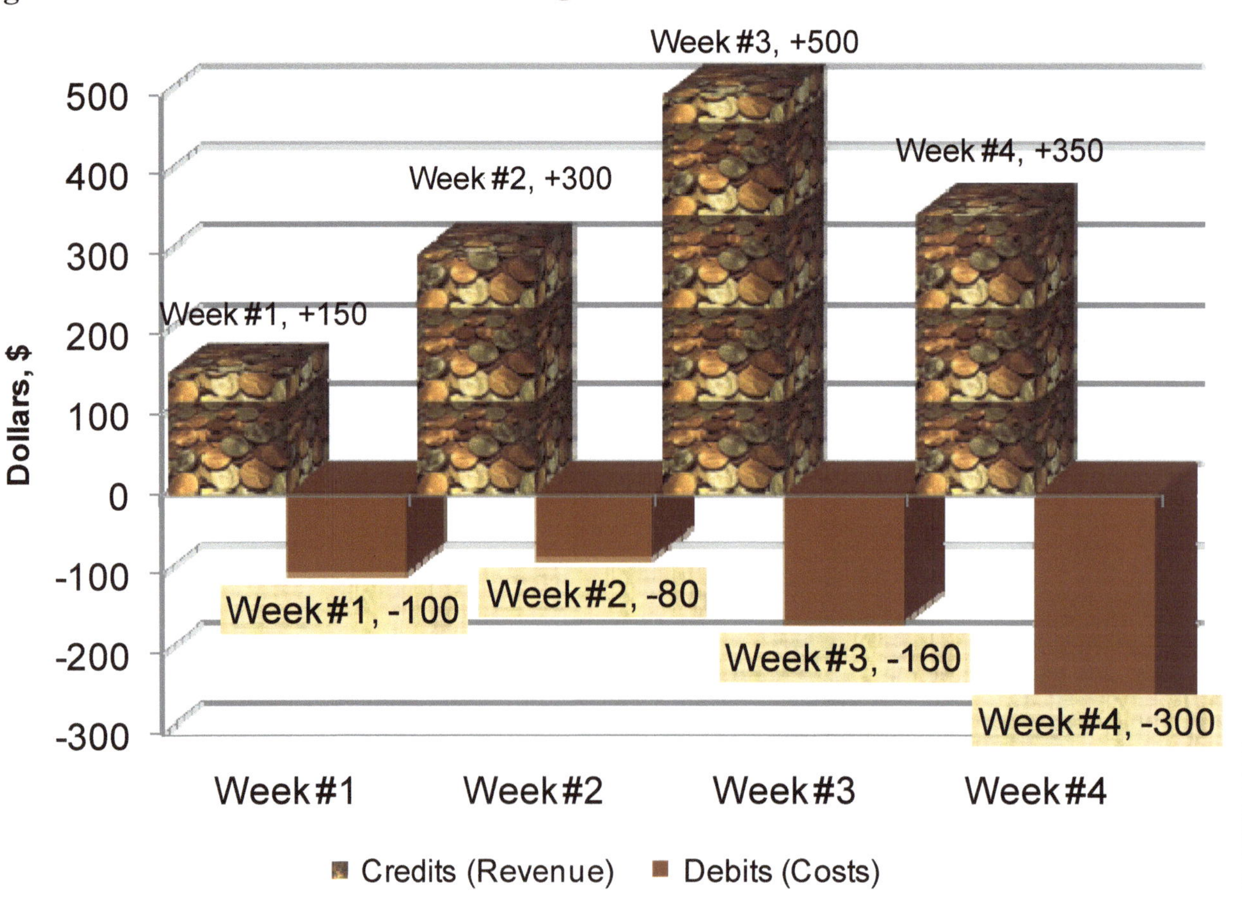

Wanda's Widgets: Month #1 Costs and Revenue

- Week #1, +150
- Week #2, +300
- Week #3, +500
- Week #4, +350
- Week #1, -100
- Week #2, -80
- Week #3, -160
- Week #4, -300

■ Credits (Revenue)   ■ Debits (Costs)

52

Workshop 1: Signed Numbers

Investigation 1.4: Adding Signed Numbers

Using Figure **1.30**, we have constructed Table 1.7 and associated column chart (Figure 1.31) that tracks Wanda's weekly account balance listing the revenue (credits), costs (debits), the sum of the weekly debits and credits, and finally the weekly checking account balance.

Assume that Wanda initially opened the new checking account with a starting balance of $1,000.00.

"*Wanda's Widgets*" Financial Tracking Table: Month #1

Table 1.7	Credits	Debits	Debits + Credits	Account Balance
				+$1000
Week #1	+150	-100	+50	+$1050
Week #2	+300	-80	+220	+$1270
Week #3	+500	-160	+340	+$1610
Week #4	+350	-300	+50	+$1660

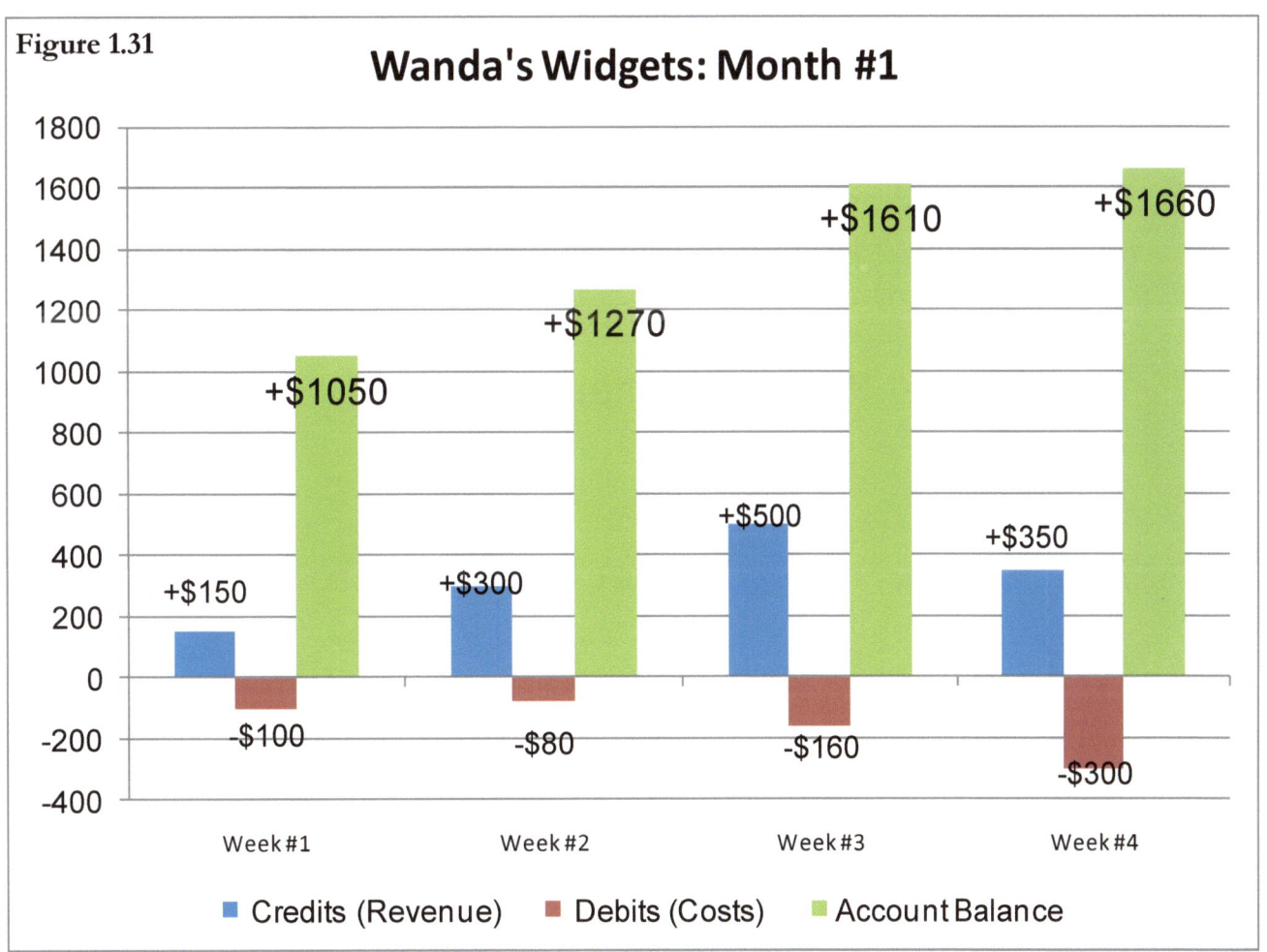

Figure 1.31

Math Toolbox

Investigation 1.4: Adding Signed Numbers

Workshop 1: Signed Numbers

Example 3: During Month #2 Wanda's widget business was set to prosper long term. She was very excited to sign large contracts with two retail establishments and felt she was on her way to becoming successful. Wanda, however, needed to expand her facility to accommodate the increased production and knew that for month #2 her costs would increase, but felt that over the long-run, this was a good business decision.

The new contracts will require Wanda to increase her costs for purchasing materials, salaries, and daily operating expenses.

1. A business makes a profit when the total revenues collected exceed the total costs incurred by the company.

    - Mathematically, this idea can be described this way:

    Profit: Revenue > Cost

2. When a business loses money, the total costs exceed the total revenue.

    - Mathematically, this idea can be described this way:

    Loss: Revenue < Cost

Math Toolbox

Workshop 1: Signed Numbers

Investigation 1.4: Adding Signed Numbers

During Month #2, Wanda's expenditures (the costs) increased as she purchased more materials, increased her production staff, and increased the week-to-week operating expenses. Figure 1.32 summarizes the revenue (credits), costs (debits) and checking account balance during Month #2.

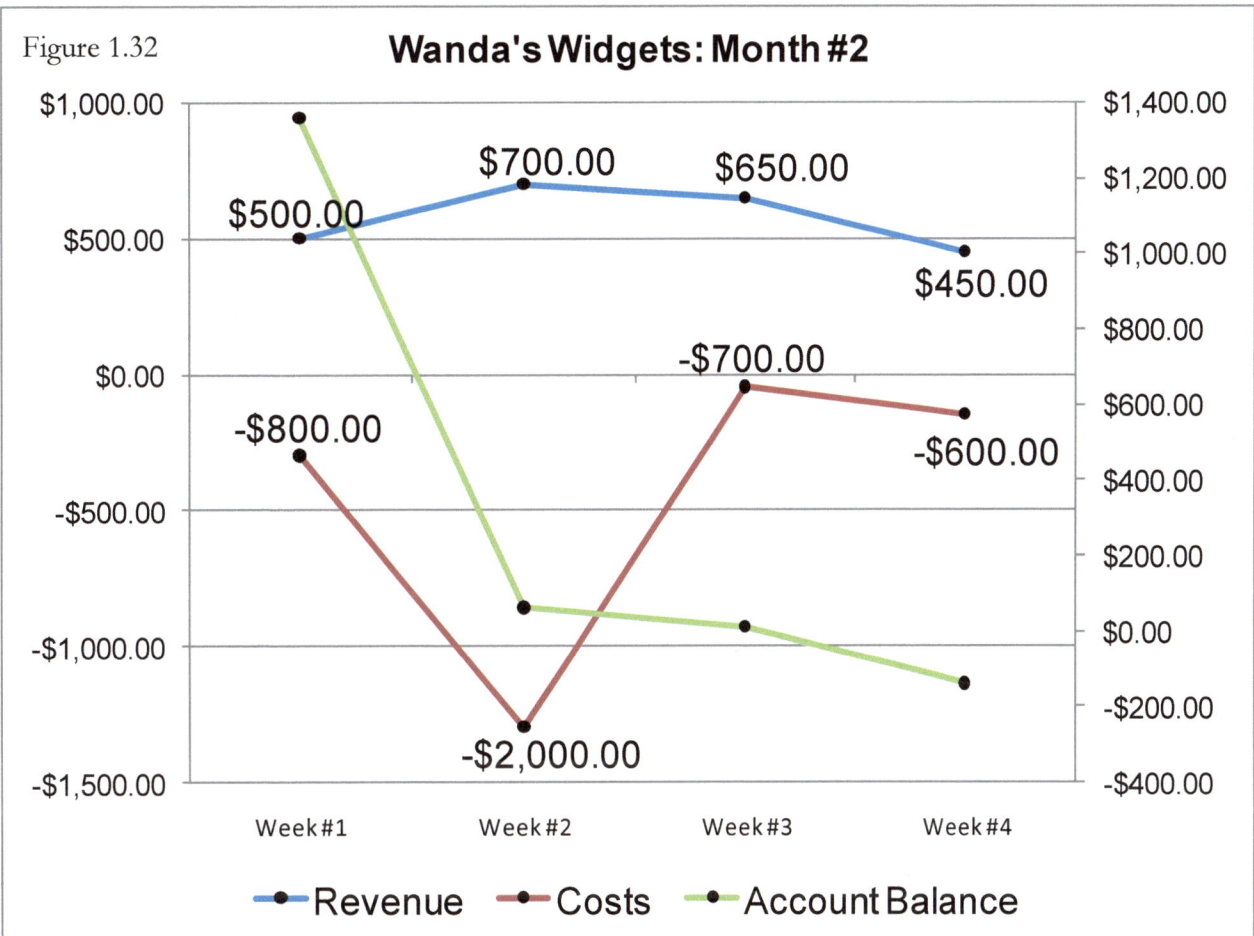

Figure 1.32

Using Figure 1.32 above answer the following questions.

1. Using the "green" Account Balance line, explain the trend in *Wanda's Widget's* account balance from Week #1 to Week #4 during the second month of operation.

2. Using the "red" Costs line above, calculate the total debits added to the checking account balance during Month #2.

3. Using the "blue" Revenue line above, calculate the total credits added to the checking account balance during Month #2.

4. Write the mathematical statement using < or > to indicate the actual relationship between the total Revenue (credits) and the total Costs (debits).

Math Toolbox

Workshop 1: Signed Numbers

Investigation 1.4: Adding Signed Numbers

At the end of Month #1, the checking account balance for ***Wanda's Widgets*** was +$1660.00. Table 1.8 below summarizes the revenue (credits), costs (debits) and checking account balance during Month #2 depicted in Figure 1.32 and 1.33.

"*Wanda's Widgets*" Financial Tracking Table: Month #2

Table 1.8	Credits	Debits	Debits + Credits	Account Balance
				+$1660
Week #1	+500	-800	-300	+$1360
Week #2	+700	-2000	-1300	+$60
Week #3	+650	-700	-50	+$10
Week #4	+450	-600	-150	-$140
Total	+2300	-4100	-1800	

Add Down ↓

Total Revenue < Total Costs: +$2300 < $4100

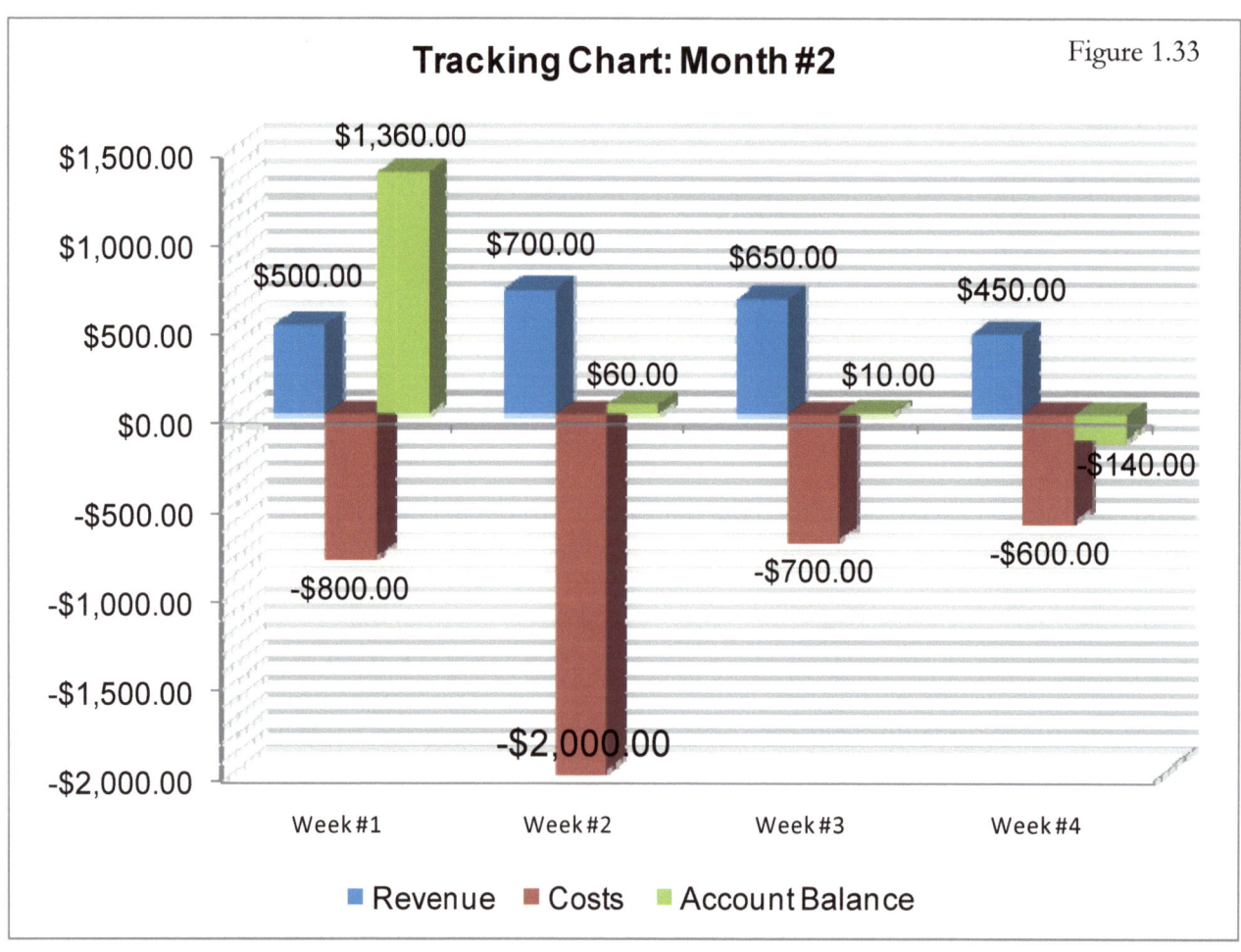

Figure 1.33

56

✎ Assignment #14
Page 1 of 2

☑ **Investigation 1.4**   Name

Instructions: During Month #3 Wanda's widget business is experiencing growing pains. Rapid growth has tapped out Wanda's cash supply, so temporarily, the business will be operating with a negative account balance. Wanda has made arrangements with the bank to guarantee the temporarily losses with equity from another source. Figure 1.34 shows the revenues collected and costs incurred during Month #3.

Use the information provided in this graph (Figure 1.34) to complete the Table 1.9 on the following page. Every blank cell in Table 1.9 must be filled with the correct number. Be sure to provide the totals requested both across (debits + credits) and down.

Figure 1.34

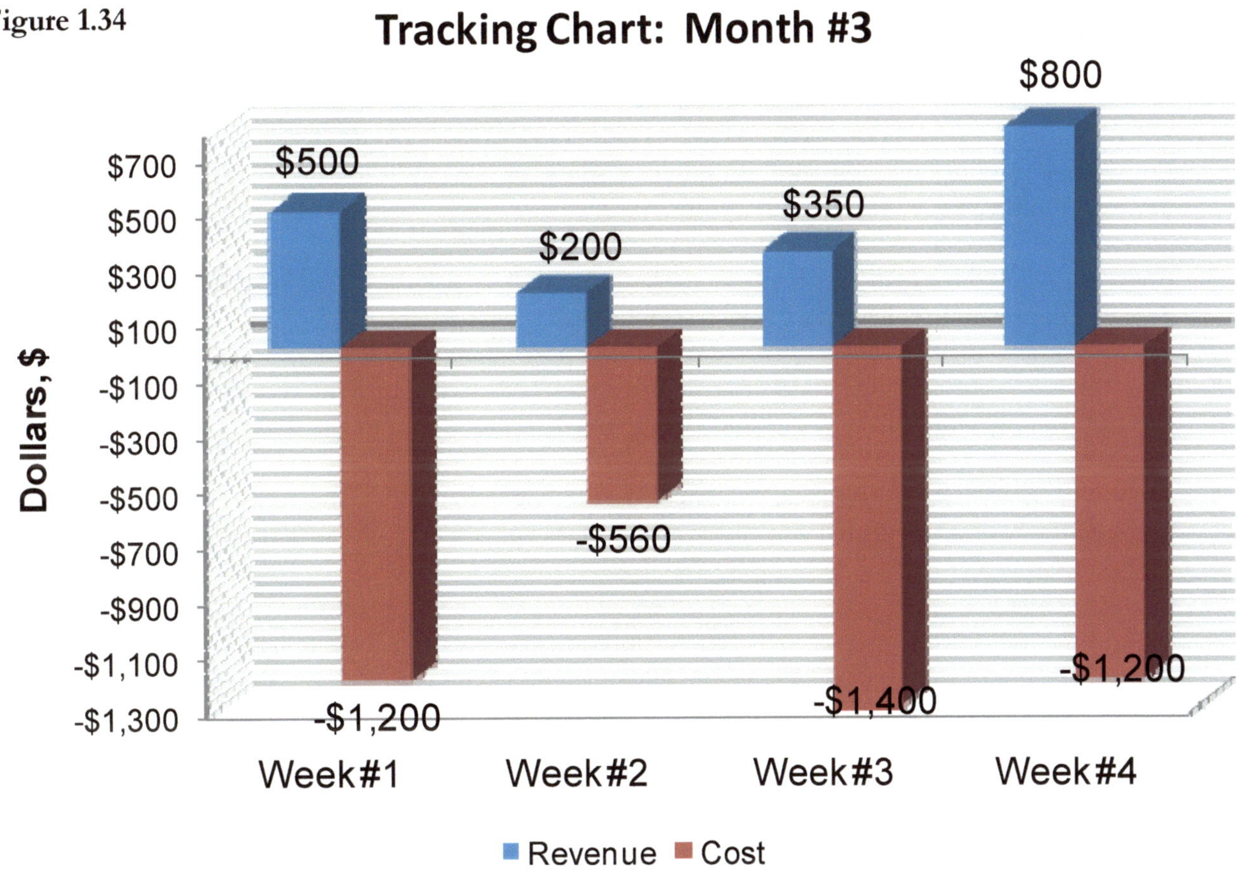

Tracking Chart: Month #3

	Week #1	Week #2	Week #3	Week #4
Revenue	$500	$200	$350	$800
Cost	-$1,200	-$560	-$1,400	-$1,200

57

✎ Assignment #14
Page 2 of 2   ☑ Investigation 1.4   Name

"*Wanda's Widgets*" Financial Tracking Table: Month #3

Table 1.9	Credits	Debits	Debits + Credits	Account Balance
				-$140
Week #1				
Week #2				
Week #3				
Week #4				
Total				

Add Down →

✎ **Once you have completed the entries for Table 1.9 above, answer the questions below. Enter your response in the boxes provided.**

1. For week #1, write a mathematical statement using < or > to indicate the relationship between the revenue collected and cost incurred.

2. For week #2, write a mathematical statement using < or > to indicate the relationship between the revenue collected and the cost incurred.

3. For week #3, write a mathematical statement using < or > to indicate the relationship between the revenue collected and the cost incurred.

4. For week #4, write a mathematical statement using < or > to indicate the relationship between the revenue collected and the cost incurred.

5. For week #1 and week #2, write a mathematical statement using < or > to compare the relationship between the actual checking account balances.

6. For week #2 and week #3, write a mathematical statement using < or > to compare the relationship between the actual checking account balances.

7. For week #3 and week #4, write a mathematical statement using < or > to compare the relationship between the actual checking account balances.

1.

2.

3.

4.

5.

6.

7.

✎ Assignment #15
Page 1 of 2   ☑ **Investigation 1.4**   Name

Instructions: In this assignment, you are provided with a partially completed financial tracking table for Month #4. Complete the table by providing totals for the "Debits + Credits" and for the "Account Balance" column. Note that the starting balance from Month #3 is shown as -$2650. Then use the relevant information from Table 1.10 to fill in the empty boxes on the column chart (Figure 1.34) found on the next page.

"Wanda's Widgets" Financial Tracking Table: Month #4

Table 1.10	Credits	Debits	Debits + Credits	Account Balance
				-$2650
Week #1	+500	-200		
Week #2	+700	-400		
Week #3	+1,000	-500		
Week #4	+1,800	-250		
Total				

1. Find 3 numbers from Table 1.10 that are smaller than −200.

2. Find 3 numbers from Table 1.10 that are larger than −500.

3. True or False. -600 > -100.

4. What is the smallest number found in Table 1.10?

5. From Table 1.10, what specifically does the following mathematical relationship refer to?   -$560 > -$1400

1.

2.

3.

4.

5.

✑ Assignment #15
Page 2 of 2   ☑ **Investigation 1.4**   Name

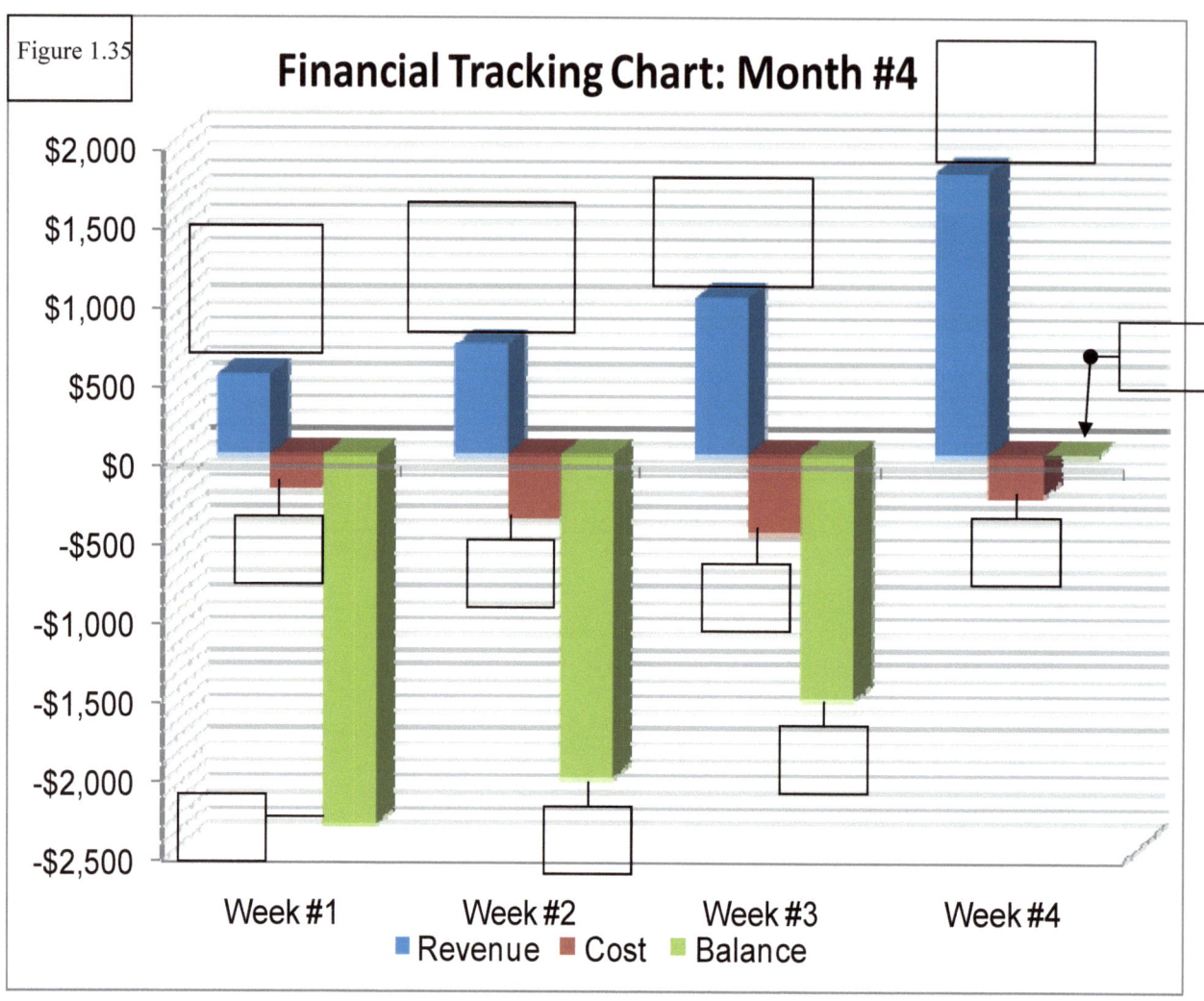

Figure 1.35

☑ Skill Builder #8   Name _____

Making a Profit

1. Tom and Ted own a business that manufactures baskets and they need to decide the minimum number of baskets to make in order that their revenue will exceed their costs. If each basket sells for $10.00, determine the minimum number of baskets needed so that Revenue > Cost.

   To find the profit, subtract the "Cost" from the "Revenue". Profit = Revenue—Cost

   (A) Cost: $256.00

   Minimum # of baskets: _____   Total Revenue: _____   Total Profit _____

   (B) Cost: $465.00

   Minimum # of baskets: _____   Total Revenue: _____   Total Profit _____

   (C) Cost: $1,550.00

   Minimum # of baskets: _____   Total Revenue: _____   Total Profit _____

   (D) Cost: $1783.00

   Minimum # of baskets: _____   Total Revenue: _____   Total Profit _____

   (E) Cost: $2,580.00

   Minimum # of baskets: _____   Total Revenue: _____   Total Profit _____

   (F) Cost: $3,005.00

   Minimum # of baskets: _____   Total Revenue: _____   Total Profit _____

   (G) Cost: $4625.00

   Minimum # of baskets: _____   Total Revenue: _____   Total Profit _____

   (H) Cost: $5372.00

   Minimum # of baskets: _____   Total Revenue: _____   Total Profit _____

☑ Skill Builder #9   Name

Using the results from Skill Builder #8, correctly label each data pair point, (# of Baskets, Profit) on the coordinate axis below. The first data pair (26 baskets, $4) is done for you.

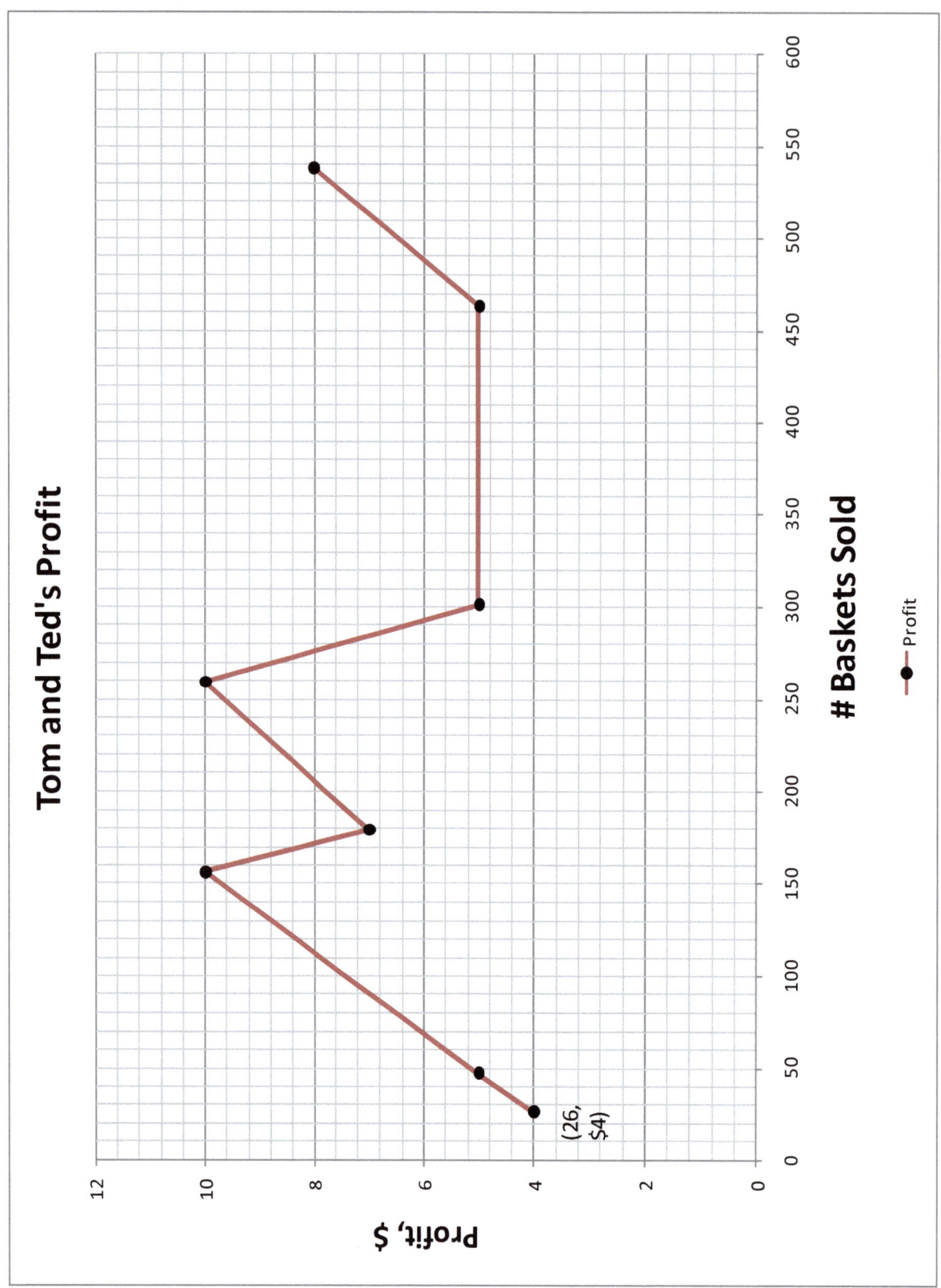

Skill Builder #10   Name

Using the results from Skill Builder #8, correctly label each data pair point, (# of Baskets, Revenue) on the coordinate axis below. One of the data pairs, (300 baskets, $3,000) is done for you.

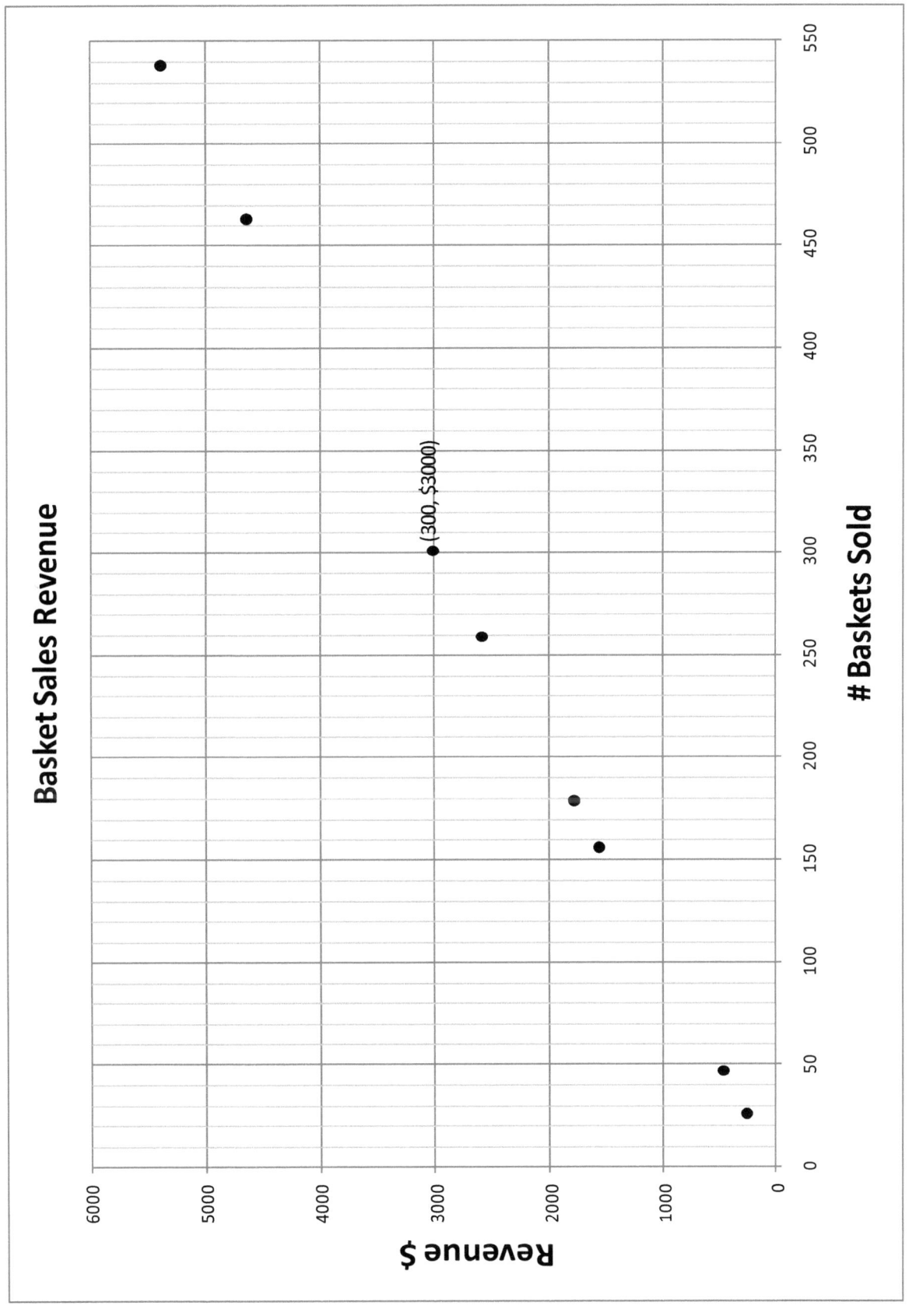

Skill Builder #11   Name

1. Using the results from Skill Builder #8, Create a table by entering the missing values below. Calculate a total for each column and enter the sum in the empty row labeled, Total.

# Baskets	Cost, $	Revenue, $	Profit, $
26		260	
47	465		5
	1550	1560	
179		1790	7
	2580	2590	10
301	3005		
463	4625		
538		5380	8
Total: 1969		19690	

2. The table below represents the "actual" number of baskets sold for 8 weeks. Use these formulas to complete the table below. The basket's sell for $10 each.

   Profit = Revenue—Cost   Profit + Cost = Revenue   Cost = Revenue—Profit

   # Baskets Sold = Revenue ÷ 10

Week	# Baskets Sold	Cost	Revenue	Profit
1	26	130		
2			300	155
3		90	150	
4	33			200
5		127		113
6		230	500	
7		99		91
8	102			560

Math Toolbox

Investigation 1.5: Calculating Change using Signed Integers

Workshop 1: Signed Numbers

Wind Chill

The term *wind chill* was first coined by Paul Siple, an Antarctic explorer when during the 1940's he and Charles Passel conducted several experiments to determine how long it took a can of water to freeze when both the temperature and wind speed were considered as joint factors. They found that the time it took the water to freeze depended on the temperature of the air and the wind speed. From these data a mathematical model was developed that could be used to predict the wind chill factor to know how cold it actually felt.

In 2001 the National Weather Service developed a new wind chill formula after a year-long cooperative effort between the United States and Canadian (www.nws.noaa.gov). The new standard is based on the wind speed at a height of 5 feet above the ground. But why, you ask, is the wind chill factor an important issue for human beings to consider?

In maintaining a constant core temperature, the human body produces energy in the form of heat. Some of this heat is radiated to the surrounding environment. Without wind an area of warm air immediately adjacent to your body acts to insulate you slightly. But as the wind blows, this insulated air next to your body is sucked away; when this happens you feel colder. How much colder you feel depends on how fast the wind moves around you.

Thus, climatologists have introduced the wind-chill factor in an attempt to measure a combination of low temperature and wind speed to study its affect on the human body.

Figure 1.36  **NWS Windchill Chart**

Wind (mph) \ Temperature (°F)	40	35	30	25	20	15	10	5	0	-5	-10	-15	-20	-25	-30	-35	-40	-45
Calm																		
5	36	31	25	19	13	7	1	-5	-11	-16	-22	-28	-34	-40	-46	-52	-57	-63
10	34	27	21	15	9	3	-4	-10	-16	-22	-28	-35	-41	-47	-53	-59	-66	-72
15	32	25	19	13	6	0	-7	-13	-19	-26	-32	-39	-45	-51	-58	-64	-71	-77
20	30	24	17	11	4	-2	-9	-15	-22	-29	-35	-42	-48	-55	-61	-68	-74	-81
25	29	23	16	9	3	-4	-11	-17	-24	-31	-37	-44	-51	-58	-64	-71	-78	-84
30	28	22	15	8	1	-5	-12	-19	-26	-33	-39	-46	-53	-60	-67	-73	-80	-87
35	28	21	14	7	0	-7	-14	-21	-27	-34	-41	-48	-55	-62	-69	-76	-82	-89
40	27	20	13	6	-1	-8	-15	-22	-29	-36	-43	-50	-57	-64	-71	-78	-84	-91
45	26	19	12	5	-2	-9	-16	-23	-30	-37	-44	-51	-58	-65	-72	-79	-86	-93
50	26	19	12	4	-3	-10	-17	-24	-31	-38	-45	-52	-60	-67	-74	-81	-88	-95
55	25	18	11	4	-3	-11	-18	-25	-32	-39	-46	-54	-61	-68	-75	-82	-89	-97
60	25	17	10	3	-4	-11	-19	-26	-33	-40	-48	-55	-62	-69	-76	-84	-91	-98

Frostbite Times: 30 minutes, 10 minutes, 5 minutes

$$\text{Wind Chill (°F)} = 35.74 + 0.6215T - 35.75(V^{0.16}) + 0.4275T(V^{0.16})$$

Where, T = Air Temperature (°F) V = Wind Speed (mph)   *Effective 11/01/01*

Math Toolbox

Workshop 1: Signed Numbers

Investigation 1.5: Calculating Change using Signed Integers

Using the Wind Chill chart in Figure 1.36, you will be introduced to several new concepts: (i) What is a variable? (ii) What happens when variables change their values? (iii) How is this change calculated?

Algebra, more than other mathematical topic, is a study in the science of patterns. When a pattern exists to describe a predictable change, variables in the form of letters are used to simplify and understand the changing process. Learning to use variable notation is a primary goal in mastering algebra.

Figure 1.36 lists a pattern of temperatures (^{0}F) across the horizontal axis (the top of the chart). To accompany the temperature, a pattern of wind speeds (in mph) is located up and down the vertical axis. Notice that both of these concepts can take on many different values and for that reason will be expressed simply by the use of two variables: T = Temperature, and W = Wind speed

1. The range in temperature extends from 40^0 F to (-45^0 F) and decreases in a predictable pattern by -5^0.

   40 + 0 = 40
   40 + (-5) = 35
   40 + (-10) = 30
   40 + (-15) = 25
   40 + (-20) = 20
   ⋮
   40 + (-85) = -45

   This kind of pattern could theoretically continue indefinitely where any integer beginning with 40 (decreasing by 5 each time) fits the pattern. We communicate this pattern for all integers by allowing a letter, such as "T", to represent all possible integer temperatures within the specified range.

   T = {40, 35, 30, 25, 20…-45}

2. The range in wind speed varies from 0 mph (calm) to 60 mph, in increments of 5 mph. So if we allow the wind speed to be represented by the variable, W, the pattern would look something like this:

   0 + 5 = 5 mph
   0 + 10 = 10 mph
   0 + 15 = 15 mph
   0 + 20 = 20 mph
   ⋮
   0 + 60 = 60

   The wind pattern (in mph) can be written this way:

   0 + W = W

   W = {0, 5, 10, 15, ...60}

3. Now that you know what is meant by a variable and you see what happens when variables change their values, the third question we need to address is: *How is this change calculated?*

In mathematics to calculate change, we use subtraction. An example of this might be helpful. Assume that you begin working a part-time job making $150 per week. After 1 year your salary has increased to $200 per week. What is the *change* in your salary? To calculate the change we subtract: $200 - $150 per week = $50 per week. The **change** in salary was $50 per week.

Math Toolbox

Workshop 1: Signed Numbers

Investigation 1.5: Calculating Change Using signed Integers

Example 1: Let us now use what was previously introduced about variables and their changing values, along with the use of subtraction to actually calculate change to see a few examples.

Using Figure 1.36, predict the temperature your body would feel for the following wind speeds and temperatures in degrees Fahrenheit.

(a) **Wind speed = 20 mph, Temperature = 30° F.** *What is the Wind Chill?*

- Step 1: Locate the wind speed of 20 mph along the vertical column.
- Step 2: Move straight across until you intersect the Temperature column (defined here as 30° F).
- Step 3: Read the wind chill temperature. (17° F) This means that your body "feels like" it is 17° F, even though the temperature is actually 30°F— much warmer than that.

What "temperature" change does your body feel due to the 20 mph wind?
(Remember, *change* means to subtract).

Symbolically we use the delta symbol, Δ, to represent change.
To calculate the change in temperature, symbolically we write, ΔT.

$$\Delta T = 17° - 30° = -13°$$

[Interpretation: Our body feels a 13 degree decrease in temperature due solely to the 20 mph wind].

- Step 4: To calculate "change", subtract the final number (30°) from the starting number (17°). In this case, the starting temperature was 30°, the final temperature was 17° F; $\Delta T = -13°$ **F.**

- Step 5: Interpret the answer. The "change" in the temperature from 30° to 17° felt by our body was a **decrease, indicated by the negative sign in, -13°F.**

(b) **Wind speed = 15 mph, Temperature = 15° F.** *What is the Wind Chill?*

- Step 1: Locate the wind speed of 15 mph along the vertical column.
- Step 2: Intersect the 15° F Temperature column.
- Step 3: Read the wind chill temperature of 0° F.
- Step 4: Calculate the "change" in temperature.
  Final Temperature—Starting Temperature = Change in temperature; $\Delta T = 0° - 15° = -15°$
- Step 5: Interpret the answer. The change in temperature from 15° to 0° is a decrease of –15° F. The human body would feel a decrease of 15° due solely to the 15 mph wind.

Workshop 1: Signed Numbers

Investigation 1.5: Calculating Change using Signed Integers

(c) Wind speed = 35 mph, Temperature = 10^0 F. *What is the Wind chill?*

- **Step 1:** Locate the wind speed of 35 mph along the vertical column.
- **Step 2:** Move straight across until you intersect the Temperature column. (defined here as 10^0 F).
- **Step 3:** Read the Wind Chill temperature of -14^0 F.
- **Step 4:** Calculate the change in Temperature, ΔT.

$$\Delta T = -14^0 - 10^0 = -24^0 F$$

- **Step 5:** Interpret your answer. The change in temperature from 10^0 to -14^0 is a decrease of -24^0 F. The human body would feel a decrease of 24^0 due solely to the 15 mph wind

(d) Wind speed = 5mph, Temperature = -5^0 F. *What is the Wind chill?*

- **Step 1:** Locate the wind speed of 5 mph along the vertical column.
- **Step 2:** Move straight across until you intersect the Temperature column. (defined here as -5^0 F).
- **Step 3:** Read the Wind Chill temperature of -16^0 F.
- **Step 4:** Calculate the "change" in temperature.
  Final Temperature—Starting Temperature = Change in temperature, ΔT.

$$\Delta T = -16^0 - (-5)^0 = -16^0 + (+5^0) = -11^0$$

> Note: Anytime it is necessary to subtract a negative value, change both the subtraction sign and the negative sign to plus signs. Subtracting a negative value is exactly the same as adding a positive value.

- **Step 5:** Interpret the answer. The change in temperature from -5^0 to -16^0 is a decrease of -11^0 F. The human body would feel a decrease of 11^0 due solely to the 5 mph wind

Math Toolbox

Workshop 1: Signed Numbers

Investigation 1.5: Calculating Change Using Signed Integers

Example 2: Notice at the bottom of Figure 1.36 are three colored boxes representing frostbite times, of 5 minutes, 10 minutes, and 30 minutes. The corresponding colors on the wind chill chart indicate that an individual could be exposed to the conditions within that color-coded region for the specified time, but after that time they would be exposed to frostbite.

Frostbite is a condition where the skin and other body tissues are damaged when exposed to extreme cold, at or below 0^0 C (32^0 F or 273 K). The blood vessels close to the skin become narrow in an attempt to preserve the core body temperature but when an individual is exposed to extreme cold for extended periods of time the protective mechanism undertaken by the body to preserve its heating mechanism can reduce the flow of blood to some areas. Eventually these areas will freeze and when this happens it is called frostbite and is most likely to happen in the parts of the body that are the farthest from the heart. The frostbitten areas will turn at first purple, then black. If not treated quickly, frostbite can cause permanent nerve damage.

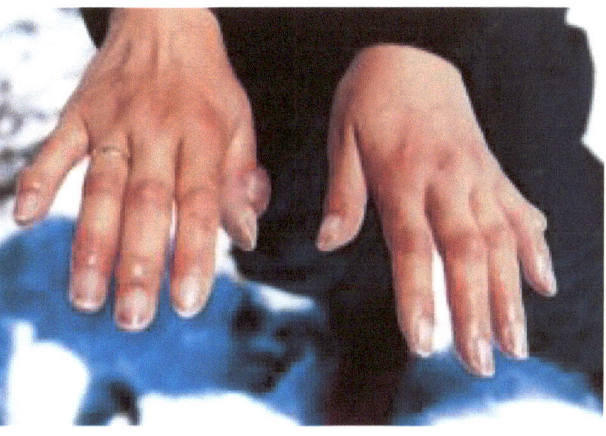

Figure 1.37 Frostbite on the Hands.

On the following page, Figure 1.38 shows a scatterplot with seven points, each is a combination of wind speed and temperature. (T, W) The respective scales are along the vertical axis—the wind speed, W—and the horizontal axis—the temperature, T.

- Each wind speed, temperature pair (T, W) correspond to a particular color coded area on figure 1.35.

  - ✓ For instance, the wind speed, temperature pair, (5^0 F, 10 mph) is found to produce a wind chill of -10^0 F located in the light blue area. This area is coded to indicate that frostbite will occur after exposure for 30 minutes.
  - ✓ The wind speed, temperature pair, (-10^0 F, 30 mph) is found to produce a wind chill of -39^0 F located in the medium blue colored area. This area is coded to indicate that frostbite will occur after exposure for 10 minutes.
  - ✓ The wind speed, temperature pair, (-20^0 F, 50 mph) is found to produce a wind chill of -60^0 F located in the purple colored area. This area is coded to indicate that frostbite will occur after exposure for just 5 minutes.

For each (T, W) pair on Figure 1.38, find the wind chill and frostbite exposure time. Enter the values in the table provided below the graph.

Investigation 1.5: Calculating Change using Signed Integers

Workshop 1: Signed Numbers

For each (W, T) pair on Figure 1.38 below, find the wind chill and frostbite exposure time from Figure 1.35. Enter the values in the table provided below the graph.

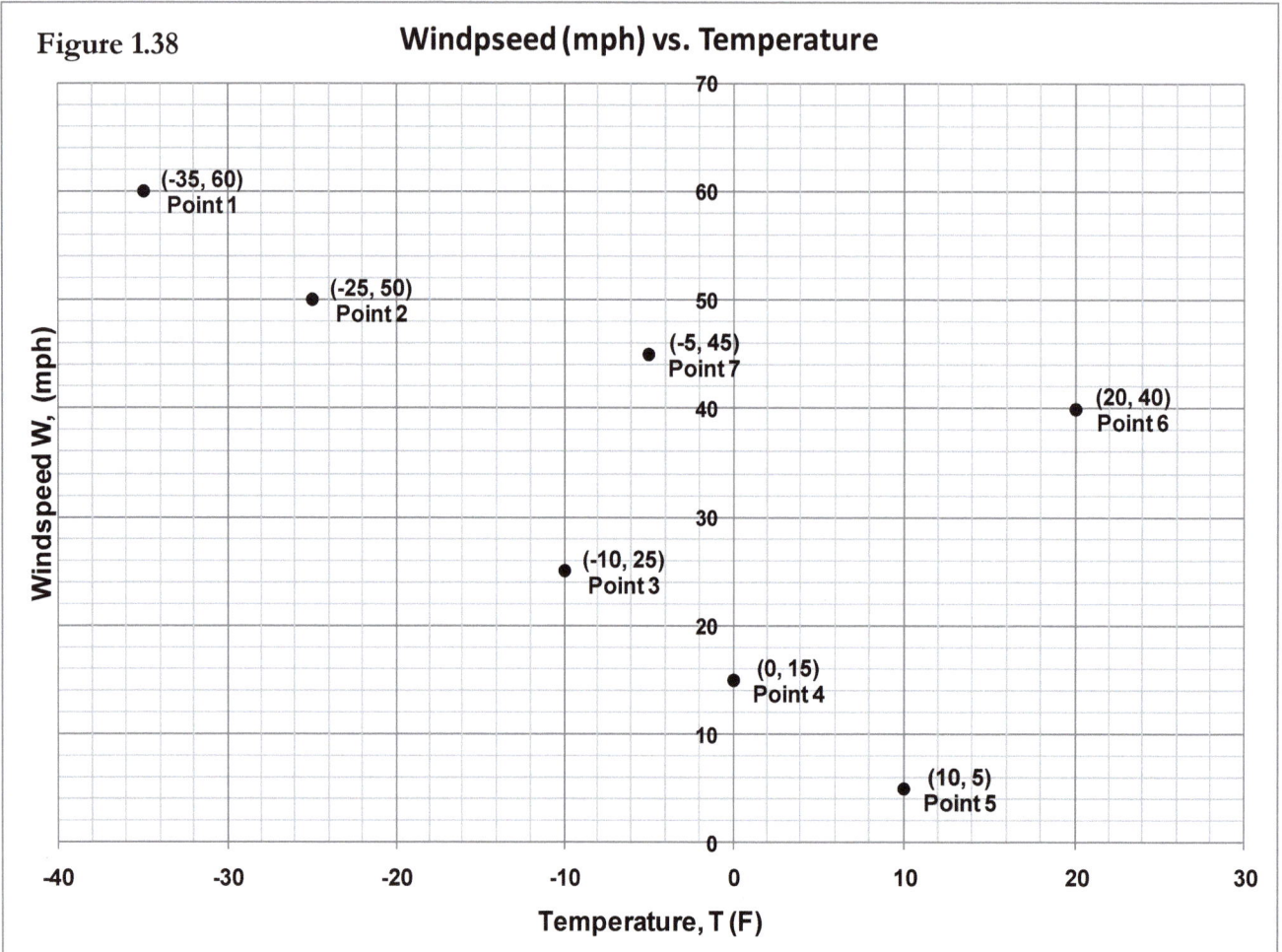

Figure 1.38

Point Number	(T, W) Pair	Wind Chill Temperature	Frostbite Exposure Time (min)	Color Coded Area
1	(-35, 60)	-84	5 min	purple
2				
3				
4				
5				
6				
7				

Math Toolbox

Workshop 1: Signed Numbers

Investigation 1.5: Calculating Change using Signed Integers

Example 3: Changes in Ocean Temperatures

Why does the ocean temperature decrease with increasing depth?

Change is evident everywhere on Earth and the ocean's temperature is no exception. Most of the heat energy of transmitted to Earth from the sun is absorbed in the first few centimeters at the ocean's surface. These surface layers heat up during the day and cool at night because this heat energy is lost to space by radiation.

Turbulence in the ocean caused by wind and waves mixes the heat near the ocean surface and eventually distribute this heat downward to deeper waters. The temperature of the surface waters varies mainly with latitude. In general the ocean temperature variation around the world is staggering. The waters around the poles corresponding to high latitude polar seas, can be as cold as $-2°$ C (about $28°$ F). The Persian Gulf on the other hand is at low latitudes and can be as warm as $36°C$ (about $97°$ F).

As the depth below sea level increases, there are places in the ocean where the temperature changes more rapidly. The graphic in Figure 1.38 shows a boundary between the surface waters (those areas of the oceans where swimming, surfing, and fishing normally take place) and the deeper layers that are not mixed with the heat from the sun. This boundary usually begins between -100 meters and -400 meters, extending several hundred meters more beneath the ocean's surface and is called the thermocline. The thermocline boundary is characterized by a rapid decrease in temperature. Figure 1.39 shows an area between -100 meters and -400 meters with a rapid temperature decrease from $22°$ C to $12°$ C.

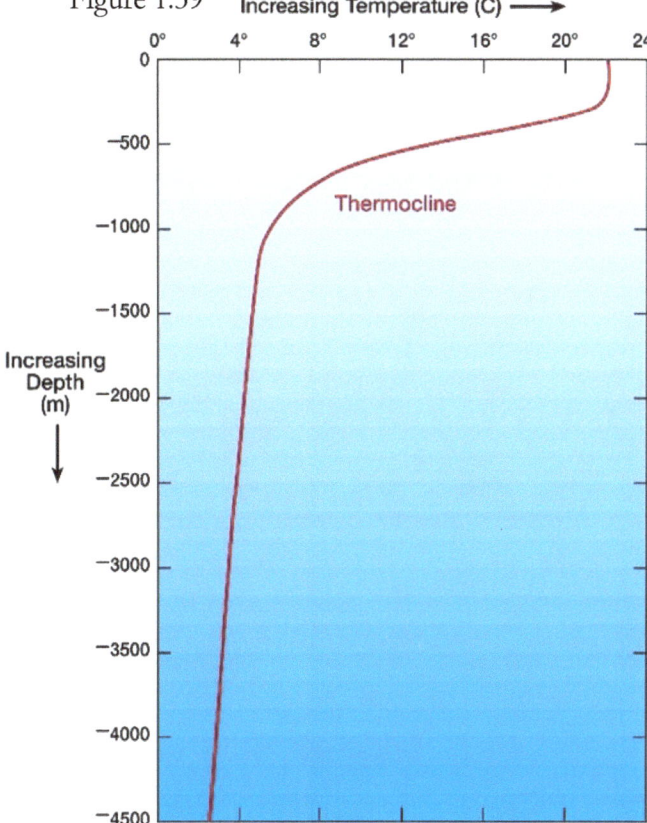

Figure 1.39

✓ What is the change in average ocean temperature, ΔT, from the polar seas to the Persian gulf?

$$\Delta T = 36 - (-2) = 36 + (+2) = 38°C$$

✓ What is the temperature change, ΔT, between -100 meters and -500 meters as described in this passage?

✓ Using the graph above, estimate the temperature of the ocean at the thermocline boundary at -2000 meters below the ocean's surface? At -4500 meters?

✒ Assignment #16
Page 1 of 2   ☑ **Investigation 1.5**   Name

Instructions: For each of the temperature and wind speed values given below, calculate the wind chill. Follow the 5 steps given in Example 1 and enter an answer in the boxes provided. Use a complete sentence when interpreting your answer and show your work when calculating ΔT. (W = Wind speed, T = Temperature)

1. W = 10 mph, T = 25° F.   Wind Chill = _____. ΔT = _____
Interpret your answer here:

2. W = 35 mph, T = 30° F.   Wind Chill = _____. ΔT = _____
Interpret your answer here:

3. W = 40 mph, T = 20° F.   Wind Chill = _____. ΔT = _____
Interpret your answer here:

4. W = 45 mph, T = -5° F   Wind Chill = _____. ΔT = _____
Interpret your answer here:

5. W = 15 mph, T = (–10°) F. Wind Chill = _____. ΔT = _____
Interpret your answer here:

6. W = 15 mph, T = (–10°) F. Wind Chill = _____. ΔT = _____
Interpret your answer here:

7. W = 10 mph, T = (–30°) F. Wind Chill = _____. ΔT = _____
Interpret your answer here:

8. W = 25 mph, T = (–15°) F. Wind Chill = _____. ΔT = _____
Interpret your answer here:

9. W = 45 mph, T = (–10°) F. Wind Chill = _____. ΔT = _____
Interpret your answer here:

✎ Assignment #16
Page 2 of 2   ☑ **Investigation 1.5**   Name

Critical Thinking Exercises:

10. Which of the two situations presented below would produce the *highest* wind chill temperature? Why?

    (a) Being exposed to a (i) wind speed of 30 mph with an air temperature of 15^0 F or (ii) Wind speed of 20 mph and a temperature of 10^0 F.

    (b) Regarding (i) and (ii) above, calculate the *change* in wind chill temperature. (show your work for credit)

11. **Challenge Question 1**: On July 21, 1983 at the Vostok research station in the Antarctic, the coldest temperature ever recorded anywhere on Earth, was -128.6^0 F. Using Figure 1.35, if exposed to this temperature, about how long would it be before an individual would experience frostbite? Express your answer using <, or >.

12. **Challenge Question 2**: Use Figure 1.36. Provide three different temperature, wind speed combinations that would produce a wind chill of (-71^0)F.

1.	2.	3.

13. **Challenge Question 3**: (a) Use figure 1.36. Show how it might be possible to experience a -58^0 F wind chill for 10 minutes without getting frostbite but in a completely different situation be exposed to frostbite after only 5 minutes at the same wind chill temperature. Explain precisely what temperatures and wind speed combinations might account for this anomaly.

(b) Given your answer to (a), calculate ΔW for the two situations.

(c) Given your answer to (a), calculate ΔT for the two situations.

☺ (d) Given your answer to (a), calculate the change in wind chill.

Assignment #17
Investigation 1.5

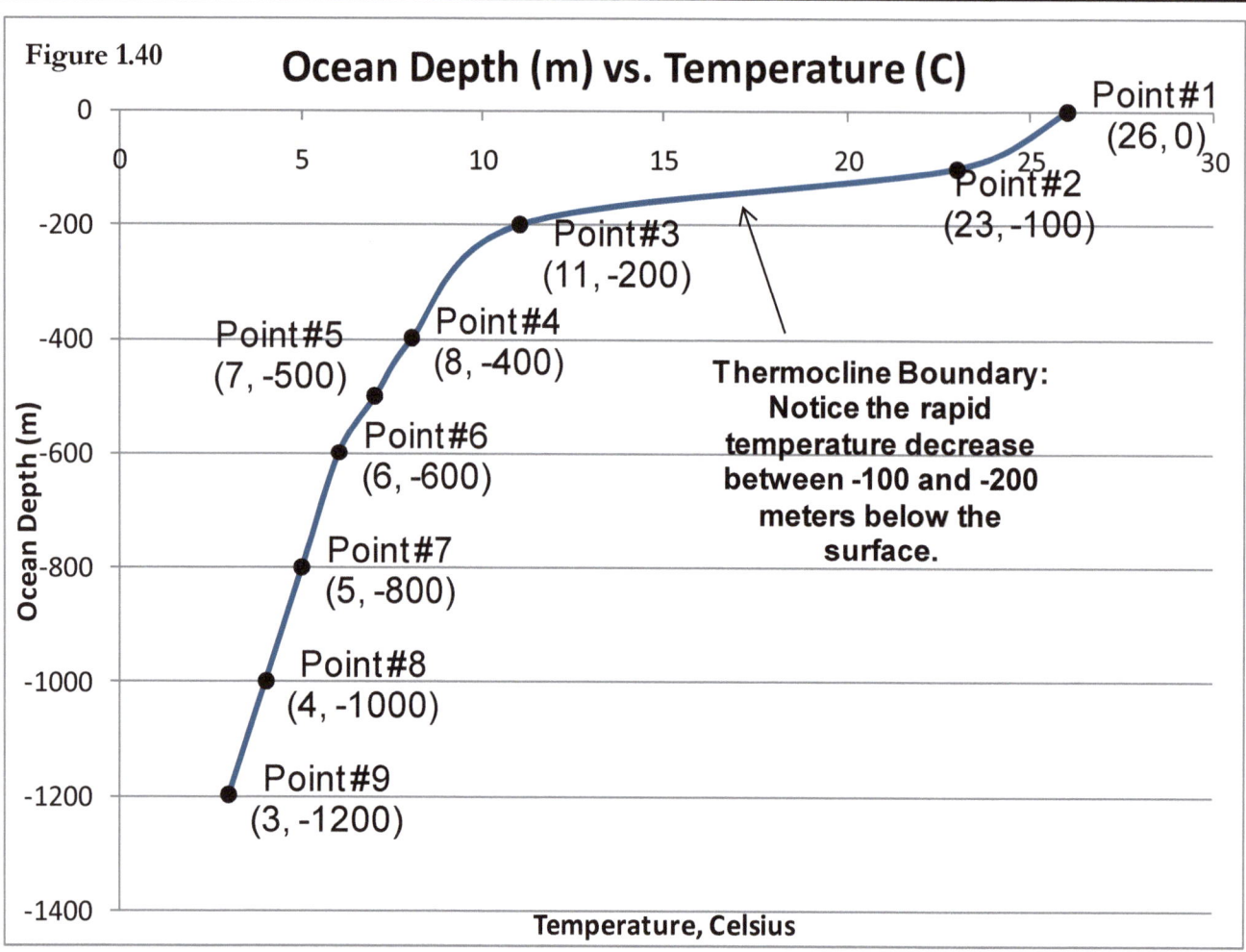

Figure 1.40 — Ocean Depth (m) vs. Temperature (C)

Thermocline Boundary: Notice the rapid temperature decrease between -100 and -200 meters below the surface.

Instructions: Using Figure 1.40 above, enter the missing values for (i) Temperature, T (ii) Depth, D in the table below.

Point	T (C)	D (meters)	Point	T (C)	D (meters)
Point #1			Point #6		
Point #2			Point #7		
Point #3			Point #8		
Point #4			Point #9		
Point #5					

✎ Assignment #17
Page 2 of 2   ☑ Investigation 1.5   Name

Instructions: Using Figure 1.40 along with the table of values on page 1 of this assignment, answer the following questions. Show your work to receive credit.

1. Calculate ΔT and ΔD between the following points:

   (a) Point #1 and Point #2. ΔT = _____ ΔD = _____

   (b) Point #2 and Point #3. ΔT = _____ ΔD = _____

   (c) Point #3 and Point #4. ΔT = _____ ΔD = _____

   (d) Point #4 and Point #5. ΔT = _____ ΔD = _____

   (e) Point #5 and Point #6. ΔT = _____ ΔD = _____

   (f) Point #6 and Point #7. ΔT = _____ ΔD = _____

   (g) Point #7 and Point #8. ΔT = _____ ΔD = _____

   (h) Point #8 and Point #9. ΔT = _____ ΔD = _____

2. Between which two points was the change in temperature the greatest?

3. Between which two points was the change in depth the greatest?

4. What is ΔT between the ocean's surface and −1200 meters?

5. What is ΔD between point #4 and point #8?

6. What is ΔD between point #3 and point #9?

7. What is ΔD between point #2 and point #6?

✎ Assignment #18
Page 1 of 1   ☑ Investigation 1.5   Name

✓ **Knowledge Check:** Read each of the supporting information for the following problems. Give your answer as an integer.

1. The coldest temperature every recorded on Earth was −129° F in Antarctica. The warmest temperature ever recorded was 136° in the Sahara Desert. Calculate the change in temperature from the warmest Temperature ever recorded on Earth to the coldest ever recorded. [Hint: You should end up with a positive integer as an answer]. Show your work for credit.

2. The coldest temperature ever recorded in the United States was −80° F in Alaska. The warmest temperature every recorded was 134° F in California. Calculate how much the temperature needed to change in going from −80° to +134° F? That is, from the coldest to the warmest ever recorded in the United States. [Hint: You should end up with a positive integer as an answer].

3. Wanda's Widget business is booming. She begins Month #3 with $1,000 in her checking account. Below is a line graph showing her checking account (debits and credit) activity for Month #3.
   Find the balance in her account at the end of week #4.

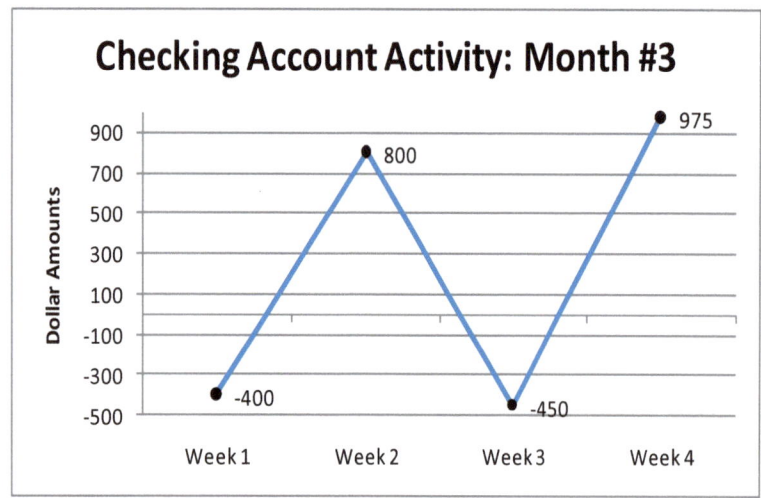

Show your work here.

4. Mt. Elbert in Colorado has a peak elevation of 14,433 feet above sea level. Consequently, the Mid-American Trench in the Pacific ocean has a maximum depth below sea level of 21,857 feet. What is the difference in elevation between these two points.

5. The average temperature on the surface of Mercury is 167° C and the average temperature on the surface of both Neptune and Uranus is −215° C. What is the change in temperature between Mercury and Uranus? [Hint: Your answer will be a positive integer].

Math Toolbox

Workshop 1: Signed Numbers

Investigation 1.6: Finding Perimeter using Algebraic Expressions

Evaluating Algebraic Expressions...What exactly does that mean?

A variable is a letter or symbol that can represent a changing set of numbers, like temperatures or wind speeds. **An expression** is a combination of variables along with signs and other numbers, that establish a pattern of many possible answers using basic arithmetic operations (addition, subtraction, etc). In this investigation we will use variables along with the arithmetic operations of "addition" and "subtraction" to create a series of answers to describe real-world objects.

Example 1: Below is a rectangle with two sides: one long side that has a length we will call "X". The short side of the rectangle has a length we will call "Y". Depending on the value given to the long and short sides (X and Y), the perimeter will vary.

What's a perimeter?

The perimeter for any geometrical figure, like a rectangle, is the total distance around the figure.

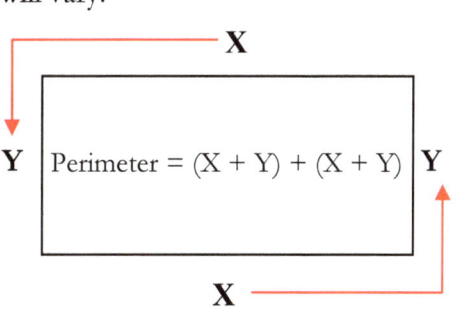

Because "X" and "Y", the lengths of the long and short sides of the rectangle, can take on different values, the perimeter can also vary. Our first goal is to create a mathematical expression that allows us to calculate the distance around the rectangle.

Using Table 1.11 below, evaluate the algebraic expression for perimeter, using the assigned length values for "X" and "Y".

Table 1.11

Length of "X"	Length of "Y"	(X + Y)	Evaluate the Expression (X + Y) + (X + Y)	Perimeter
10 m	5 m	(10 + 5) = 15	(10 + 5) + (10 + 5) = 15 + 15	30 m
4 m	1 m	(4 + 1) = 5	(4 + 1) + (4 + 1) = 5 + 5	10 m
100 ft	50 ft	(100 + 50) = 150	(100 + 50) + (100 + 50) = 150 + 150	300 ft
25 in	10 in	(25 + 10) = 35	(25 + 10) + (25 + 10) = 35 + 35	70 in
12 miles	4 miles	(12 + 4) = 16	(12 + 4) + (12 + 4) = 16 + 16	32 miles
5 km	1 km	(5 + 1) = 6	(5 + 1) + (5 + 1) = 6 + 6	12 km

Math Toolbox

Workshop 1: Signed Numbers

Investigation 1.6: Finding Perimeter using Algebraic Expressions

Example 2: The figure below is called an "isosceles trapezoid". Two sides of this shape are the same length and are identified with the variable, X. The other two sides are different lengths. "Y" represents the length of the third side, "Z" represent the length of the fourth side.

Using the table below, evaluate the algebraic expression for perimeter, using the assigned length values for "X", "Y", and "Z"

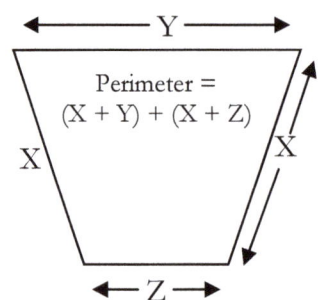

Table 1.12

Length of "X"	Length of "Y"	Length of "Z"	Evaluate the Expression (X + Y) + (X + Z)	Perimeter
10 m	5 m	3 m	(10 + 5) + (10 + 3) = 15 + 13	28 m
4 m	2 m	1 m	(4 + 2) + (4 + 1) = 6 + 5	11 m
100 ft	50 ft	25 ft	(100 + 50) + (100 + 25) = 150 + 125	375 ft
25 in	10 in	5 in	(25 + 10) + (25 + 5) = 35 + 30	65 in
12 miles	4 miles	2 miles	(12 + 4) + (12 + 2) = 16 + 14	30 miles
5 km	3 km	1 km	(5 + 3) + (5 + 1) = 8 + 6	14 km

Example 3: An equilateral triangle is a three sided shape where all sides are equal in length. Since all three sides are of equal length, we use the same variable, X, to denote the identical length.

The perimeter is found by adding (X + X + X).

But another, more resourceful way to express this same concept is to multiply "X" by 3.

Perimeter = (X + X + X) = 3X

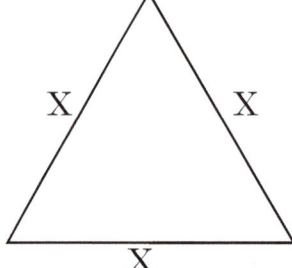

Length of X	Evaluate the Expression (3X)	Perimeter	Length of X	Evaluate the Expression (3X)	Perimeter
5 in	(3)(5)	15 in	10 ft	(3)(10)	30 ft
20 km	(3)(20)	60 km	12 cm	(3)(12)	36 cm
4 meters	(3)(4)	12 m	15 miles	(3)(15)	45 miles

Math Toolbox

Workshop 1: Signed Numbers

Investigation 1.6: Finding Perimeter using Algebraic Expressions

Example 4: Two new shopping malls are being planned in separate cities. One of the shopping malls will be built in the shape of a rectangle, the other in the shape of a trapezoid. The mall designers are planning to include a walking path around the perimeter of each for shoppers to not only shop, but to exercise. The figures below show the dimensions and shape of the perimeter walking paths. (a) Which of these two shopping malls will have the greatest distance for customer exercise? Let "A" represent the perimeter of Shopping mall A, and "B" represent the perimeter of Shopping mall B. Write your final answer as an expression using < or > to compare the perimeters of the two malls.

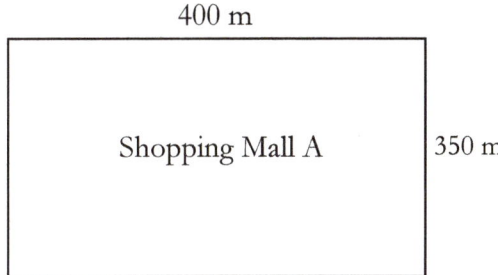
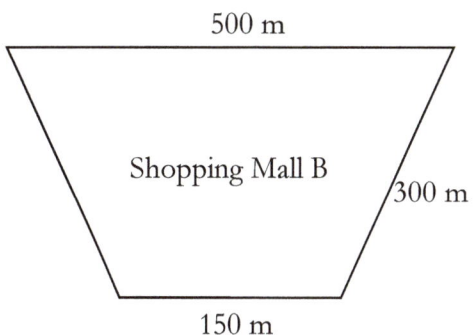

Perimeter of Shopping Mall A: (400 m + 350 m) + (400 m + 350 m) = (750 m) + (750 m) = 1500 m.

Perimeter of Shopping Mall B: (300 m + 500 m) + (300 m + 150 m) = (800 m) + (450 m) = 1250 m

Because the perimeter of Shopping Mall A is greater than the perimeter of Shopping Mall B (1500 m > 1250 m), write the final answer as an inequality:
A > B.

(b) What is the *difference* in perimeter distance of the two shopping malls?

The word "difference" in this question is a key word that means "to subtract".

> The difference in perimeter distance for the two shopping malls is: 1500 m—1250 m = 250 m

(c) What is the *total* perimeter distances planned for both shopping malls?

The word "total" in this question is a key word that means "to add".

> The total perimeter distance for both shopping malls is 1500 m + 1250 m = 2750 m.

Math Toolbox

Workshop 1: Signed Numbers

Investigation 1.6: Finding Perimeter using Algebraic Expressions

Example 5: When a diagonal line is drawn from corner to corner of a square, two right triangles are produced. They are called *right triangles* because each contains a 90^0 angle denoted by the small square in the corner.

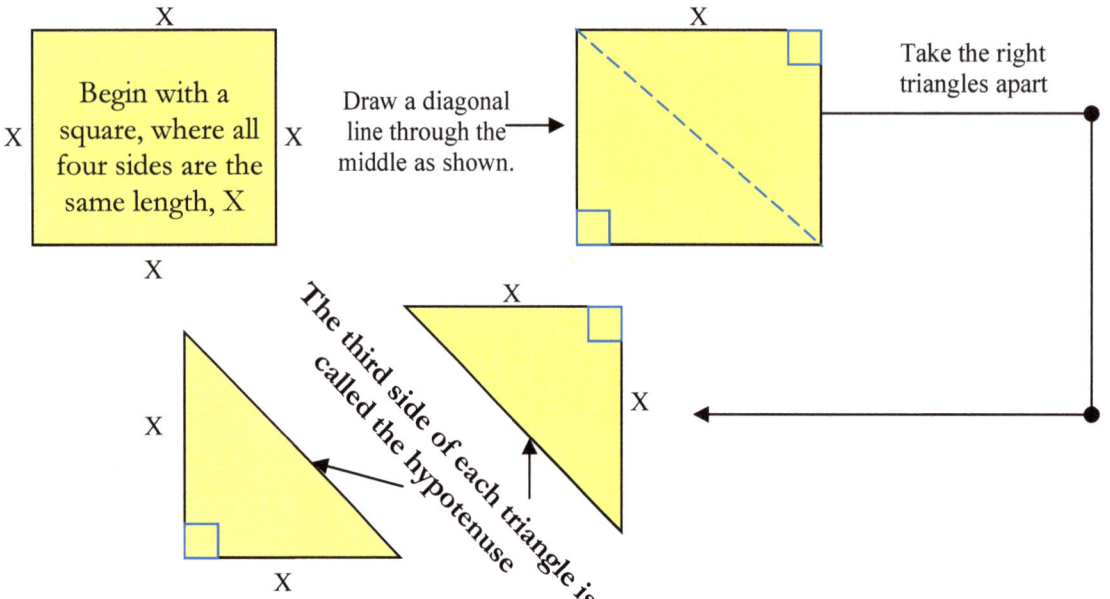

The word geometry is formed from two Greek words:

- 🌐 The first is *geo*, that means "earth".
- ■ The second is *metron*, that means "to measure".

Literally, the word geometry means to *measure the Earth* and although we don't literally measure the Earth with geometry, understanding the mathematical principles behind these and other geometrical shapes are used to solve real-world problems.

Assume the square shown above is measured such that each side has a length of 5 inches.
   The perimeter, then, would be calculated as :
- ■ Perimeter = (X + X + X + X) = 4X = (4)(5) = 20 inches.

⊗ What you might be tempted to say at this point is that since the square was divided into the two right triangles, each right triangle has a perimeter that is half of this 20 inches. **But that would wrong!**

Look again at the perimeter which is the distance around the outside of the figure, not through the middle using the dashed line as a pathway. These and other wrong assumptions are easy to make. We will make you aware of these pitfalls as topics like this are introduced.

Math Toolbox

Workshop 1: Signed Numbers

Investigation 1.6: Finding Perimeter using Algebraic Expressions

Example 6: When a diagonal line is drawn from corner to corner of a rectangle, two right triangles are produced because each contains a 90^0 angle, the length of the sides are different such that: $X > Y$.

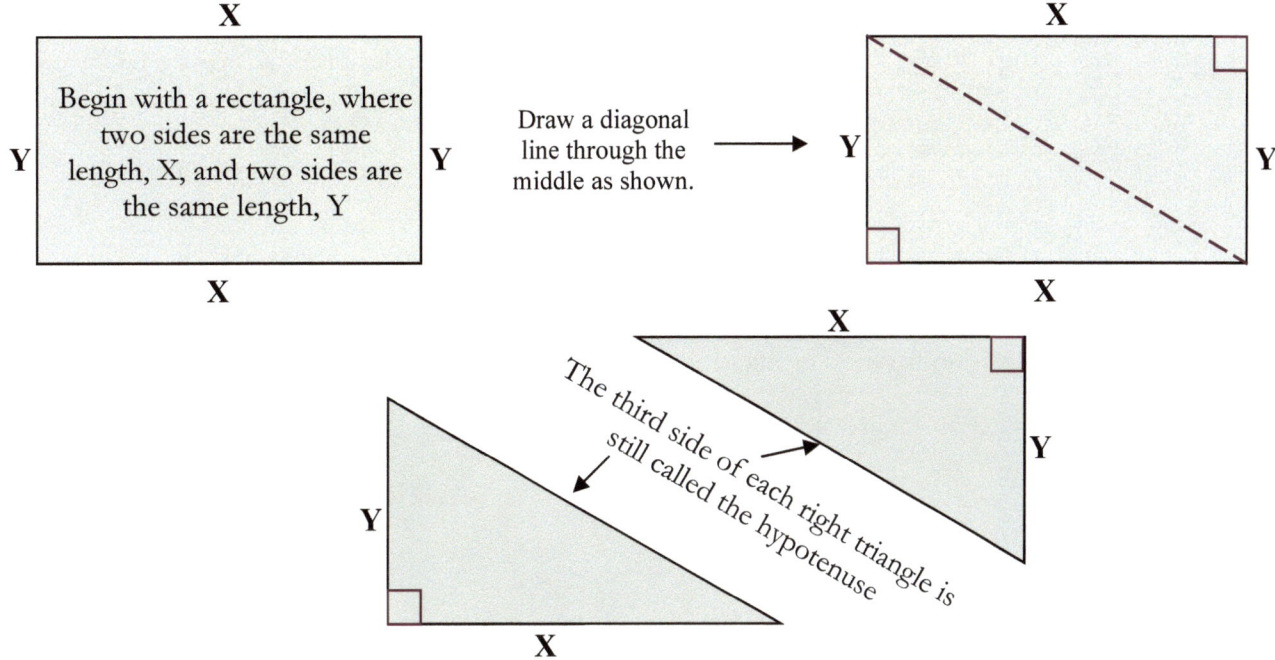

(a) Given the rectangle above, assume that $X = 3$ units long and $Y = 4$ units long. Find the perimeter by evaluating the algebraic expression: $(X + Y) + (X + Y)$.

✓ The perimeter of the rectangle is $(3 + 4) + (3 + 4) = 7 + 7 = 14$ units around.

(b) If "X" and "Y" are still 3 and 4 units respectively, and the perimeter of the right triangles are each 12 units, can you determine the length of the hypotenuse? Call the length of the hypotenuse, Z.

We do this by "solving" a simple algebraic equation: $X + Y + Z = 12$

❖ Remember that the perimeter is calculated by adding the length of all the sides together.

But we know the length of side X. $X = 4$.
We know the length of side Y. $Y = 3$.
The only missing piece is the length of side Z. $Z = ?$

Let's replace the variables in the equation to find the perimeter with their known values.
$4 + 3 + Z = 12$ (Simplify by adding together the known values)
$7 + Z = 12$ (What number added to 7 gives us 12?)
$Z = 5$ (Of course the answer is 5).

✎ Assignment #19
Page 1 of 3   ☑ Investigation 1.6   Name

Instructions: Find the perimeter for the following geometrical shapes. Remember to include the unit of length when giving your answer. Without the unit of length, your answer will be counted wrong.

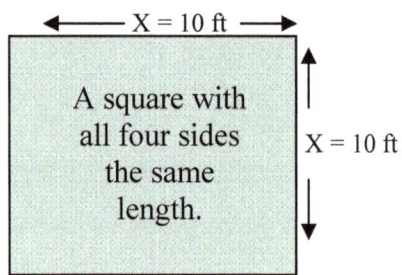

A square with all four sides the same length. X = 10 ft

1. Write the algebraic expression for perimeter.

2. Plug in the correct values for "X" shown left.

3. Simplify your expression, showing the perimeter.

---

4. Write the algebraic expression for the Trapezoid perimeter.

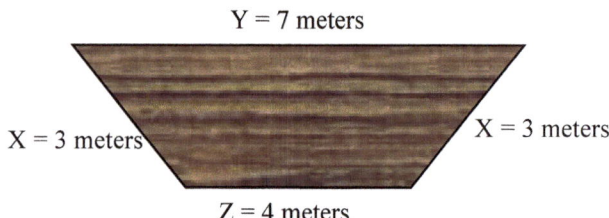

5. Plug in the correct values for "X, Y, Z" shown Right.

6. Simplify your expression showing the perimeter.

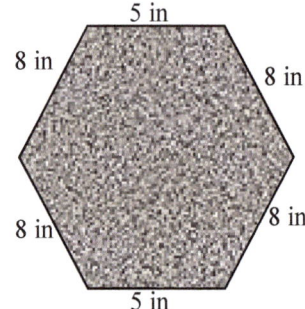

7. How many sides does this shape have? _____

8. How many **_different_** lengths are there? _____

9. How many **_different_** variables must be used to represent the perimeter (Hint: consider your answer to #8). _____

10. Create an algebraic expression that you could use to calculate the perimeter.

11. Plug in the values shown. Calculate the perimeter.

✎ Assignment #19 ☑ **Investigation 1.6**   Name
Page 2 of 3

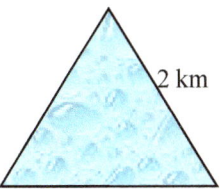

The shape shown at the left is an equilateral triangle (don't confuse this with a right triangle).

12. How many sides does this shape have? _____

13. How many *different* lengths are there? _____

14. How many *different* variables must be used to represent the perimeter (Hint: consider your answer to #13). _____

15. Create an algebraic expression that you could use to calculate the perimeter.

16. Plug in the values shown. Calculate the perimeter.

The Stop sign at the right is unique in that all sides equal 12 inches in length.

17. How many sides does this shape have? _____

18. How many *different* lengths are there? _____

19. How many *different* variables must be used to represent the perimeter (Hint: consider your answer to #18). _____

20. Create an algebraic expression that you could use to calculate the perimeter.

21. Plug in the values shown. Calculate the perimeter.

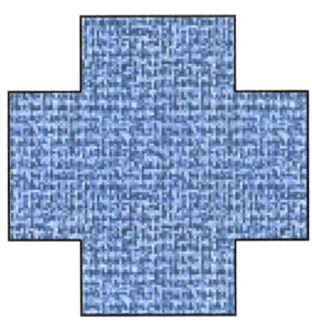

22. **Challenge Question!**

The shape at the left requires only two variables to determine its perimeter. Based on this fact and using "X" and "Y" as your variables, write an algebraic expression that can be used to calculate the perimeter of this shape.

For the two values given: X = 10 cm and Y = 3 cm, calculate the perimeter by using your algebraic expression.

Assignment #19
Page 3 of 3
Investigation 1.6
Name

24. Find the perimeter of a square table top if each side is 3 feet long. _____

25. A five-sided figure has sides of length 5 ft, 3 ft, 2 ft, 8 ft, and 12 ft. Find the perimeter. _____

26. How much fence is required to enclose a rectangular yard that is 220 ft by 100 ft? _____

27. Calculate the perimeter of the top of a square CD case if the length of one side is 7 inches. _____

28. A backyard is shaped like the trapezoid shown below.

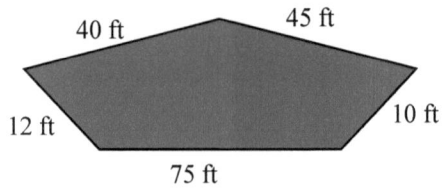

How much fence is needed to enclose the entire yard? _____

29. The perimeter of a square box lid measures 48 inches. What is the length of one side?

30. A room with two identical closets (shown on either end) is being built.

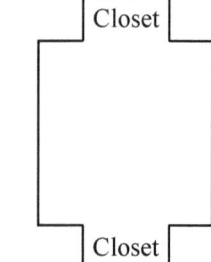

(a) If only three variables are needed to calculate the perimeter of the room, what algebraic expression would allow this calculation?

(b) How many longest sides (assuming they are equal in length) are there?

(c) How many shortest sides (assuming they are equal in length) are there?

(d) How many other sides (assuming they are equal in length) are there?

(e) If the Longest sides measure 15 ft each, the shortest sides measure 5 feet each, and the perimeter is 86 feet, what is the length of the remaining two sides (assuming they are equal in length).

Show your work below.

Assignment #20
Page 1 of 1

☑ **Investigation 1.6**   Name

Integrated Review:

Perform the following integer operations and be sure to indicate your final answer as a positive (+) or a negative (-) value.

1. (+100) + (-10) + (-30) = _____

2. (-100) + (-10) + (-30) = _____

3. (+150) + (250) — (-500) = _____

4. (-15) + (-20) — (+30) + (15) = _____

5. (+200) + (-25) + (-200) = _____

6. (+2) + (-4) + (+5) + (-11) = _____

7. (-80) — (+30) + (+50) = _____

8. (5) + (-4) + (-6) — (-10) = _____

9. Your current bank balance is (-$50). On Friday you deposit $100 in cash, and on that day an automatic debit of (-$60) has been deducted from your account.

What is your balance on Friday? _____

...However, several weeks ago you notified the health club to cancel your membership and on Monday morning you ask the health club to credit your account $60. This was done on Monday.

What is your balance on Monday? _____

10. True or False: Subtracting a debit is the same mathematical operation as adding a credit.

11. True or False: when adding two or more positive numbers, the result will never be a positive number.

12. True or False: Subtracting a negative number is the same as adding a positive number.

13. Evaluate the expression for the given values: X = 10, Y = 20, Z = 5

    X — Y + Z = _____

14. Evaluate the expression for the given values: a = -10, b = -20

    a — b = _____

15. Evaluate the expression for the given values: a = +20, b = -30

    a + b = _____

16. A SCUBA diver descends from the ocean's surface a distance of 100 feet under water. Finding an interesting cadre of fish on a coral reef the diver remains there for some time. After about 30 minutes, the diver descends to the next underwater site that lies 60 feet below his current position.

How far from the ocean's surface is the diver? _____

17. How many units from zero, and in which direction does the number –14 lie?

18. Beginning at zero on the number line you first move to the right 5 units. After this you move to the left 10 units and then back to the right 15 units. Where are you on the number line?

Math Toolbox

Investigation 1.7: Multiplying Signed Integers

Workshop 1: Signed Numbers

Falling Bodies

Remember Aristotle? He was the Greek philosopher (384 B.C.—322 B.C.) who influenced scientific thought for some 2000 years. He developed theories on nature and physics that are completely different from what is understood today. The reason Aristotle is important is because of the enormous influence he had on even the greatest thinkers throughout history. Galileo is one of those.

Galileo Galilei (1564—1642)

One question that Aristotle posed to later generations is "What causes motion?". Galileo, in turn was concerned with *how* things move rather than *why* they move. Galileo showed that when objects fall, regardless of size, mass, or shape, all objects accelerate equally. This concept is strictly true if we are able to neglect air resistance, or air friction, when objects are in free fall. Remember Galileo's famous demonstration of this concept from the leaning tower of Pisa.

As legend would have it, Galileo dropped a stone and a feather from the leaning tower of Pisa and surmised that they accelerate at the same rate. This acceleration is due to gravity. The theory that heavier objects fall faster directly opposes this idea. We will see here why Aristotle was wrong, and Galileo was right.

The difference between Aristotle's theory—that heavier objects fall faster—requires that gravity be selective. It is because of gravity that objects fall in the first place. If two objects are dropped from some height and are in freefall, gravity accelerates them in the same way. Their shape may act to increase friction or air flow around them, but when friction is ignored and gravity is the only thing acting on them, their speeds will be the same.

Figure 1.41 Leaning Tower of Pisa

Every planet and every moon in the solar system has a different gravitational pull due to the particular planetary mass. When the pull of gravity is strong, objects accelerate faster and attain a faster speed over time. For instance, objects fall faster on Earth than they do on Moon. Why? Because the mass of the Earth is greater. The greater the planet's mass, the greater the gravitational pull.

In a later assignment, you will be asked to imagine Galileo's experiment on other planets in the solar system. The objects falling from the tower would necessarily fall at different speeds due to differences in gravity. As the pull of gravity gets stronger, the objects fall faster.

Math Toolbox

Workshop 1: Signed Numbers

Investigation 1.7: Multiplying Signed Integers

So what is gravity?

Gravity accelerates objects, and the extent to which that acceleration is measured, is the value that is assigned as the gravitational constant. For instance, on Earth, objects change their speed (accelerate) at a rate of 10 meters per second, every second. Since this pull is downward, it customary to assign it a negative value. (we use the variable "g" to represent gravity).

Therefore, the acceleration due to gravity on Earth is written this way: $g = -10 \dfrac{m}{\sec^2}$

The negative sign indicates that gravity pulls downward! Have you ever dropped a book and had it accelerate upward? Of course not! So when dealing with gravity's pull by any planet in the solar system, we use the negative sign to indicate that it pulls *downward*.

The next question to consider is what happens to the speed of an object when it falls freely? Does the speed increase, decrease, or stay the same? Imagine dropping a ball from a 10th story window in a high rise. As the ball accelerates toward Earth, it's speed changes, because when we speak of acceleration that's what the word means. Accelerating down the highway means that the speed of the car is increased. So, overtime, when gravity pulls on an object that falls freely through the air, it's speed accelerates. The longer it falls, the faster it falls, downward.

This is where "g" comes into play. If a ball is dropped from the window of a 10-story building that stands about 300 meters tall, the ball accelerates by changing its speed 10 meters per second, each second that it falls to Earth. The ball continues to accelerate downward until it hits the ground.

⌛ After one second the ball is traveling -10 m/sec. After two seconds the ball is traveling –20 m/sec. After three seconds the ball is traveling –30 m/sec, and so on. The gift that Galileo gave to the world was in understanding the answer to the question, "How does the speed change?". The answer to that question is according to the *pull* of gravity.

The table below summarizes the mathematical constant representing gravitational pull for each of the planets in our solar system. (These numbers are rounded as integers to make the multiplication easier).

Table 1.13

Planet	Mercury	Venus	Mars	Jupiter	Saturn	Uranus	Neptune	Pluto
Gravitational pull (m/sec^2)	-4	-9	-4	-25	-10	-8	-12	<1

Math Toolbox
Workshop 1: Signed Numbers
Investigation 1.7: Multiplying Integers

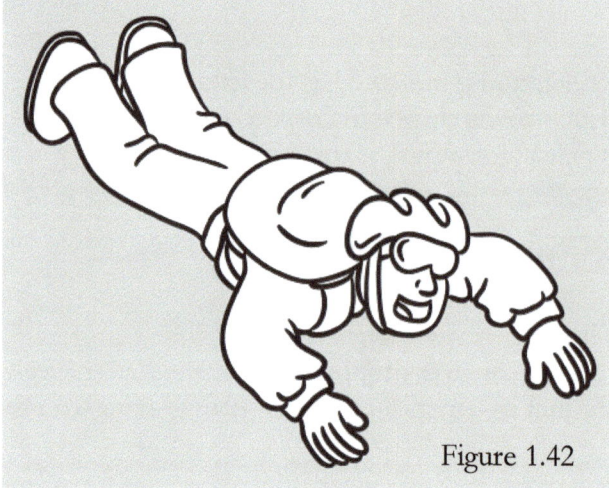

Figure 1.42

Example 1 A skydiver is in free fall through the Earth's atmosphere having jumped from an aircraft flying 3,000 meters (about 9,000 feet) overhead. The table below shows the speed of the skydiver as it increases from the time he leaves the aircraft (speed = zero) to 1,000 meters above the surface when it is time to deploy the main parachute.

In an actual skydive, the speed of the skydiver would reach a maximum value called "terminal velocity". This is due to atmospheric frictional effects. When the downward acceleration due to gravity is affectively neutralized by the upward force of drag due to the size, mass, and other variables, a maximum speed is reached.

✓ To calculate the speed of the skydiver before he reaches terminal velocity, multiply the time (in seconds) by the gravitational constant. On Earth, g = -10. When multiplying a positive value (like time) by a negative value (like the acceleration due to gravity), **the result is a negative number**. In this case, the negative number is speed—and because the skydiver continues to fall "downward", the speed must be a negative number.

Table 1.14

Time in Freefall	*gravity* × *time*	Speed, S
5 seconds	$(-10) \times 5 = -50$	-50 m/sec
7 seconds	$(-10) \times 7 = -70$	-70 m/sec
8 seconds	$(-10) \times 8 = -80$	-80 m/sec
9 seconds	$(-10) \times 9 = -90$	-90 m/sec
10 seconds	$(-10) \times 10 = -100$	-100 m/sec
11 seconds	$(-10) \times 11 = -110$	-110 m/sec

✓ **The speed of the skydiver is dependent on both the total time in freefall and the acceleration due to gravity.**

Challenge Question: The acceleration due to gravity on the moon is one-sixth that of Earth. Would the speed downward change faster or slower on the moon than it would on Earth?

Math Toolbox

Workshop 1: Signed Numbers

Investigation 1.7: Multiplying Signed Integers

Example 2. In mathematics just like graphs are used as a shorthand way to display a picture of the data, symbols can also be used to present broad concepts in shorthand. Recall that the speeds in Example 1 were calculated beginning 5 seconds after the skydiver jumped from the aircraft and ending 11 seconds after jumping from the aircraft. It is much easier if we assign variable letters to each of the changing values.

What values are changing? (i) Time is changing and (ii) Speed is Changing.

- To make the notation easier to write about, let time be represented by the variable, t. Let speed be represented by the variable, s.
- Because the times we use begin with 5 seconds and end with 11 seconds, it customary to write this symbolically using the symbols below. The < sign is now enhanced by a line underneath. This is read "less than or equal to".
- Interpreting the symbolism, the inequality below is read this way: "time, t, is greater than or equal to 5 seconds, but less than or equal to 11 seconds". This takes in all integer time values between 5 seconds and 11 seconds and includes the 5 and 11 second times.

$$5 \leq t \leq 11$$

Suppose the times we used range from 1 second to 20 seconds. How would this be written symbolically?

$$1 \leq t \leq 20$$

- Interpreting the symbols used above, the inequality is read this way: "time, t, is greater than or equal to 1 second, but less than or equal to 20 seconds". This inequality then takes in all integer time values between 1 second and 11 seconds.
  (This includes the values 1 second and 20 seconds).

A long distance truck driver reports to work at 6 a.m. on Monday morning and drives for 12 hours, until 6 p.m. that night. If we assume that 6 a.m. begins his day, (t = 0), then 6 p.m. ends his day 12 hours later (t = 12). Write this range of values for "t" has an inequality.

$$0 \leq t \leq 12$$

Investigation 1.7: Multiplying Signed Integers

Workshop 1: Signed Numbers

Example 3

Just like using symbolic language allows us to say a lot about ranges in data, a graph also allows us to create a broad picture of data. The graph below summarizes the data table showing the change in speeds for the skydiver between 5 and 11 seconds.

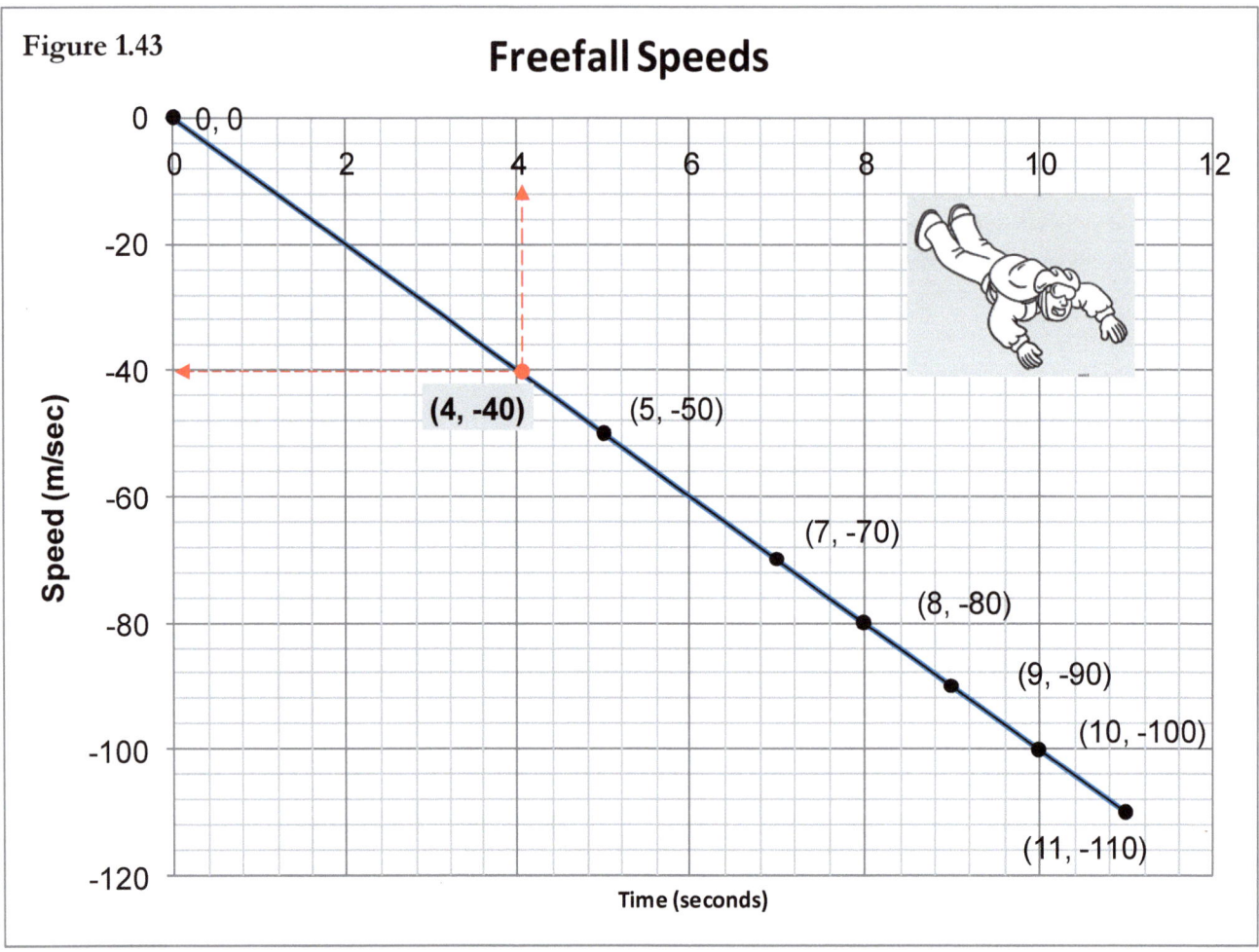

Figure 1.43

✓ One benefit of a graph is that it can be used to make predictions about some event that isn't specifically addressed in the data table used to generate the graph. One example of this is shown by the red dashed line that descends from the time, t = 4, down to the actual graphed line joining the points. Then following the red line to the left we see that it intersects the speed, S = -40 m/sec.

In this way, we can make a **prediction** about the skydiver's speed after 4 seconds in freefall. According to the graph his speed will be –40 m/sec after being in freefall for 4 seconds.

Investigation 1.7: Multiplying Signed Integers

Workshop 1: Signed Numbers

Example 4: As you have seen, mathematics uses the negative sign to indicate many different things. A negative checking account balance, indicated by -$100, means the account is overdrawn. A negative temperature, -40⁰ F means the temperature is below zero and thus is very cold. The negative sign can also indicate direction, like when considering speeds, the negative sign can indicate the speed is happening in a downward direction. Speaking only mathematically a "negative speed" is less than zero. But in the real world, a "negative speed" indicates direction. To solve this problem in mathematics, it is often necessary to express the **absolute value** of a number. This is the distance the number lies from zero on the number line. The symbol for the absolute value is two parallel lines: | |

$|-3|$ Means, "How far from zero is the number –3 on the number line?" The answer, of course, is 3 units. We see the negative sign, in this case, indicates direction. The number, -3, lies to the left of zero 3 units. So we say, the absolute value of –3 = 3.

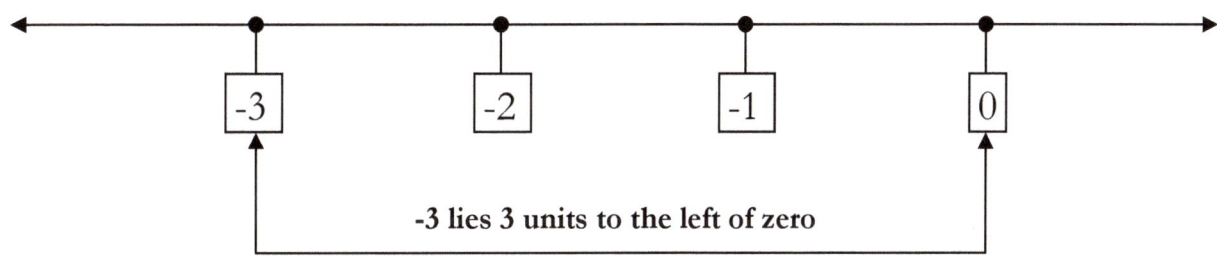

-3 lies 3 units to the left of zero

Since the absolute value of a number is that number's *distance* from zero, the absolute value of a number is always positive, or 0. The absolute value is *never* negative.

Consider the following examples:

$|-2| = 2$ **Interpretation:** The number –2 lies two units to the left of zero on the number line.

$|-5| = 5$ **Interpretation:** The number –5 lies 5 units to the left of zero on the number line.

$|-100| = 100$ **Interpretation:** The number –100 lies 100 units to the left of zero on the number line.

$|-8-(-2)| = |-8+(+2)| = |-6| = 6$ Simplifying the expression inside the absolute value bars first, gives a result inside the absolute value bars of –6. This number then simplifies to +6.

Assignment #21
Page 1 of 2
☑ **Investigation 1.7**   Name

Instructions: Using the data table presented in Example 1 below showing the gravitational constants for the planets in our Solar System, complete a data table for a range of times, then answer the questions posed.

Table 1.13

Planet	Mercury	Venus	Mars	Jupiter	Saturn	Uranus	Neptune	Pluto
Gravitational pull (m/sec^2)	-4	-9	-4	-25	-10	-8	-12	<1

Figure 1.43

Mercury has been described by some as a baked planetary stone where life as we know it would be unbearable. Because Mercury is one of the smallest planets in the solar system, its gravitational pull is quite small when compared to Jupiter or Saturn (or Earth for that matter).

Figure 1.43 shows an artists conception of the interplanetary probe, "Messenger" in orbit around Mercury that is set to begin in 2011.

Mercury's surface temperature varies, ranging from a chilly (–457° F) to (–438° F). As it orbits the Sun, its distance also varies between about 29 million miles to about 44 million miles.

1. Using the variable "T" to represent temperature, write an inequality showing that the temperature on Mercury ranges from –457° F to –438° F.

2. Using the variable "D" to represent distance, write an inequality showing that the distance Mercury lies from the sun varies between about 29 million miles to about 44 million miles.

3. Using the values of gravity found in Table 1.13 above, write inequality expressions that correctly express the relationship among the various values of "g". For instance: -1 > -5. Enter your answers in the table below.

Planet Name	Planet Name	Gravity: Inequality Expression	Explanation
Mercury	Venus	-4 > -9	The gravitational pull on Mercury, g = -4, is stronger than that on Venus, g = -9.
Venus	Jupiter		
Jupiter	Saturn		
Saturn	Neptune		

Assignment #21
Investigation 1.7

4. Assuming that the *Messenger Probe* is able to establish orbit around Mercury, **calculate the speed** a dropped temperature probe would attain for the given times if it fell through Mercury's atmosphere toward the planet's surface. Use the value for "g" from the Table 1.13. Place your answers in the table below.

Time in Freefall	Gravity x Time =	Speed, S (m/s)
5 seconds	(-4) x (5) =	-20
6 seconds		
7 seconds		
8 seconds		
9 seconds		
10 seconds		
11 seconds		
20 seconds		

The Messenger spacecraft is the first one designed to orbit Mercury—the planet closest to the sun. It flew past Mercury on January 14, 2008, and made the first up-close measurements since Mariner 10's final flyby in 1975.

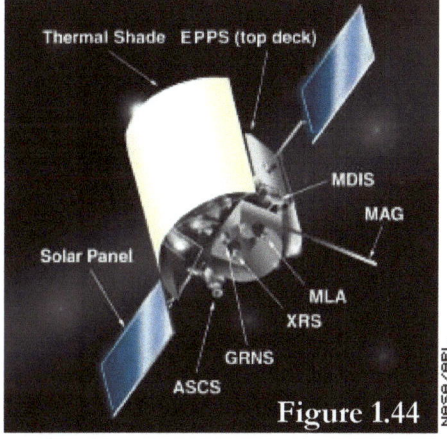

Figure 1.44

According to NASA, "Scientists have argued about the origins of Mercury's smooth plains and the source of its magnetic field for more than 30 years. Now, analyses of data from the January 2008 flyby of the planet by the Mercury Surface, Space Environment, Geochemistry and Ranging (MESSENGER) spacecraft have shown that volcanoes were involved in plains formation and suggest that its magnetic field is actively produced in the planet's core. The controversy over the origin of Mercury's smooth plains began with the 1972 Apollo 16 moon mission, which suggested that some lunar plains came from material that was ejected by large impacts and then formed smooth "ponds." When Mariner 10 imaged similar formations on Mercury in 1975, some scientists believed that the same processes were at work. Others thought Mercury's plains material came from erupted lavas, but the absence of volcanic vents or other volcanic features in images from that mission prevented a consensus."

5. Using the values calculated from #4, complete the table below.

Time (sec)	Speed, m/s	Absolute value	Time (sec)	Speed, m/sec	Absolute Value
t = 5	-20	$|-20| = 20$	t = 9		
t = 6			t = 10		
t = 7			t = 11		
t = 8			t = 20		

Assignment #22 — Investigation 1.7

On May 28, 2008, Michael Fournier planned to make history by setting a world freefall record and steal the glory from one Joseph Kittinger. After leaping from a balloon, Fournier believed he would break the 760mph sound barrier within 37 seconds. The lack of friction due to the thinness of the air would have meant he could attain a much higher terminal velocity, and his team of scientists estimated that he would reach a top speed of 1,113mph before he is slowed down by greater air resistance. His parachute was planned to deploy after six minutes, 25 seconds. He would reach the Earth after approximately eight and a half minutes. Before Fournier, American Joseph Kittinger jumped from 31,333 meters as part of a medical experiment. Also, in 1962 Russian jumper Evgueni Andreiev jumped from 24,483 meters to set a world freefall record.

Figure 1.45

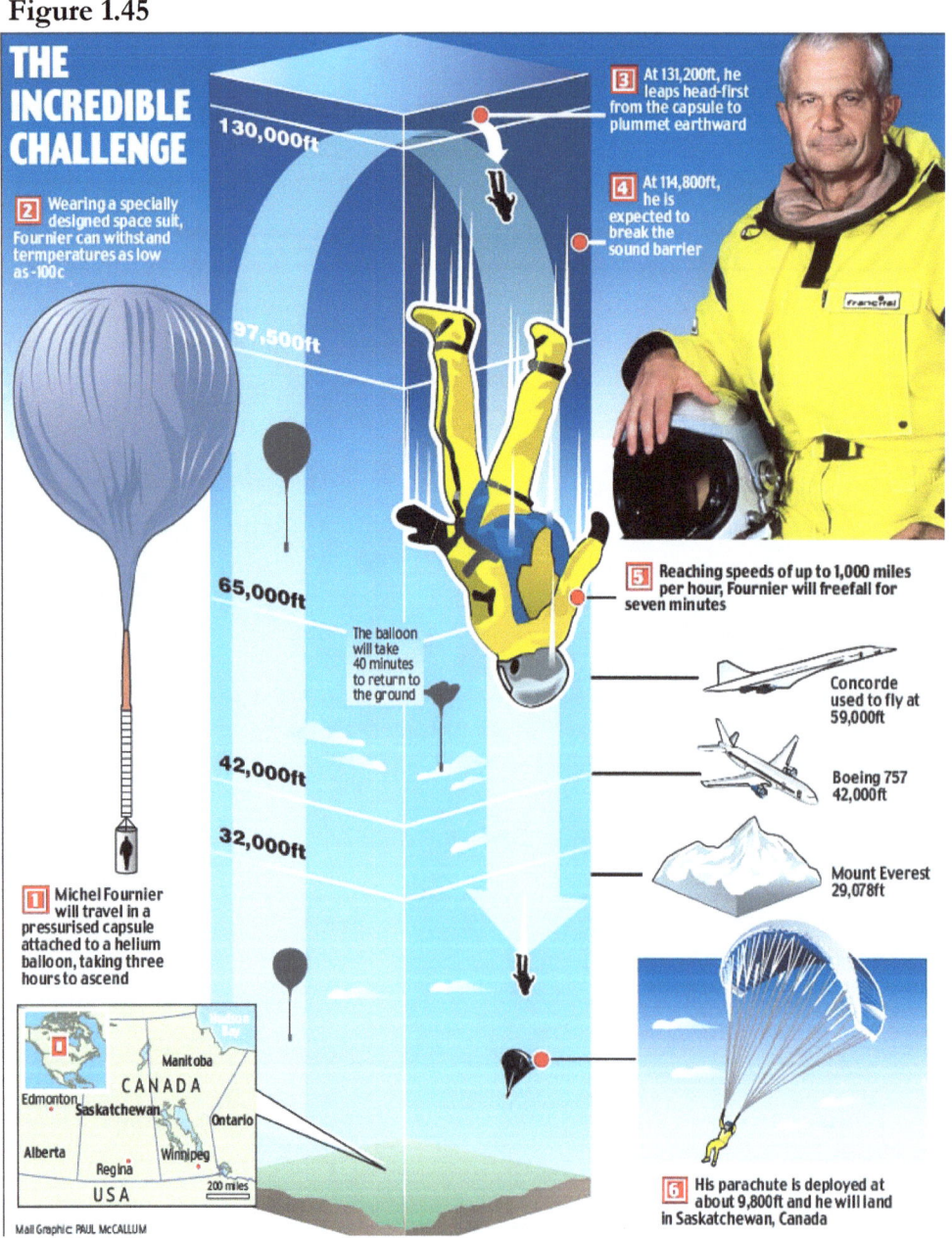

Dubbed the *incredible challenge*, Michael Fournier plans to wear a specially designed space suit that will allow him to withstand temperatures as low as -100^0 C. Wearing it he will jump from a height of 131, 200 feet allowing gravity to accelerate him to speeds that will eventually exceed the sound barrier.

Gravity will accelerate the would-be hero changing his speed by 32 feet per second, each second he is in freefall. At 114,800 feet, he is expected to exceed the speed of sound. Eventually, he will reach speeds > 1,000 mph. An estimate indicates that he will reach a speed of about 1,113 mph after having been in freefall for a little over 6 minutes. His parachute will deploy about 9,800 feet above sea level.

The acceleration due to gravity can be measured using meters or feet.
$g = -10$ m/sec^2
$g = -32$ ft/sec^2.

✎ Assignment #22 ☑ **Investigation 1.7** Name
Page 2 of 3

After reading the additional information regarding Figure 1.45, answer the following questions or provide the appropriate calculations.

1. The acceleration due to gravity in freefall can be measured using feet rather than meters. If $g = -32$ ft/sec^2, complete the following table regarding Michael Fournier's planned freefall from the balloon. Assume he exits the balloon with a speed, $S = 0$.

Table 1.15

Time (seconds)	Speed, S (m/s) *gravity • time*	Absolute Value
t = 0	0	0
t = 2	$-32 \cdot 2 = -64$	$|-64| = 64$
t = 3		
t = 4		
t = 5		
t = 6		
t = 7		
t = 10		

2. Write an inequality expression that describes the following speeds.

Example: If we consider the time between t = 0 and t = 2 seconds, write a corresponding expression for speed.

$0 \leq t \leq 2$ This means that for the freefall times between t = 0 and t = 2 seconds (including the values, 0 and 2), the corresponding expression for speed is:

$$-64 \leq S \leq 0$$

Interpretation: As the time in freefall varies between t = 0 and t = 2 seconds, the speed changes from 0 to –64 meters per second. Remember that the "-" sign here indicates direction, i.e., downward. The expressions, however, must both be mathematically correct.

✎ Assignment #22
Page 3 of 3   ☑ **Investigation 1.7**   Name

Table 1.16

Time Range	Speed (Inequality) Expression	Interpretation
$2 \leq t \leq 3$		
$3 \leq t \leq 4$		
$4 \leq t \leq 5$		
$5 \leq t \leq 6$		
$6 \leq t \leq 7$		
$7 \leq t \leq 10$		

3. Calculate the change in speed as requested in the following questions.

   Example 1: What is the *change in speed* between t = 0 and t = 2 seconds, $0 \leq t \leq 2$?
   The speed at zero seconds is 0 and the speed at 2 seconds is –64.
   To calculate *change in speed*, we subtract:
   $$-64 - 0 = -64$$

   Example 2: If the speed at t = 1 is –32 m/s and the speed at t = 2 is –64 m/s, what is the change in speed for the interval: $1 \leq t \leq 2$

   To calculate the *change in speed*, we subtract: $-64 - (-32) = -64 + (+32) = -32$

   Calculate the *change in speed* for the following time intervals? (Show your work for credit)

   (a) $2 \leq t \leq 3$

   (b) $3 \leq t \leq 4$

   (c) $4 \leq t \leq 5$

   (d) $5 \leq t \leq 6$

   (e) $6 \leq t \leq 7$

Mars! Is it possible that life once existed there? NASA has sought an answer to this important question for decades.

Figure 1.46

Figure 1.46, a mosaic of Mars, is the view you might see from a spacecraft orbiting the "Red Planet". The center of the photographic scene shows the entire Valles Marineris canyon system that extends over 4,000 km or about 2486 miles across the surface. It's depth is estimated to be 7 kilometers below the surface, or about 5 miles deep. Compared to Valles Marineris, the Grand Canyon on Earth seems quite small at 446 km or about 277 miles long. The Grand Canyon extends about 1.6 km (or about 1 mile) deep into the Earth's surface.

Three volcanoes are visible as dark red spots attaining elevations of 18 to 26 km above the planet's surface. (about 11 to 16 miles above the surface of Mars). The temperature on the surface of Mars varies between 150 K and 310 K.

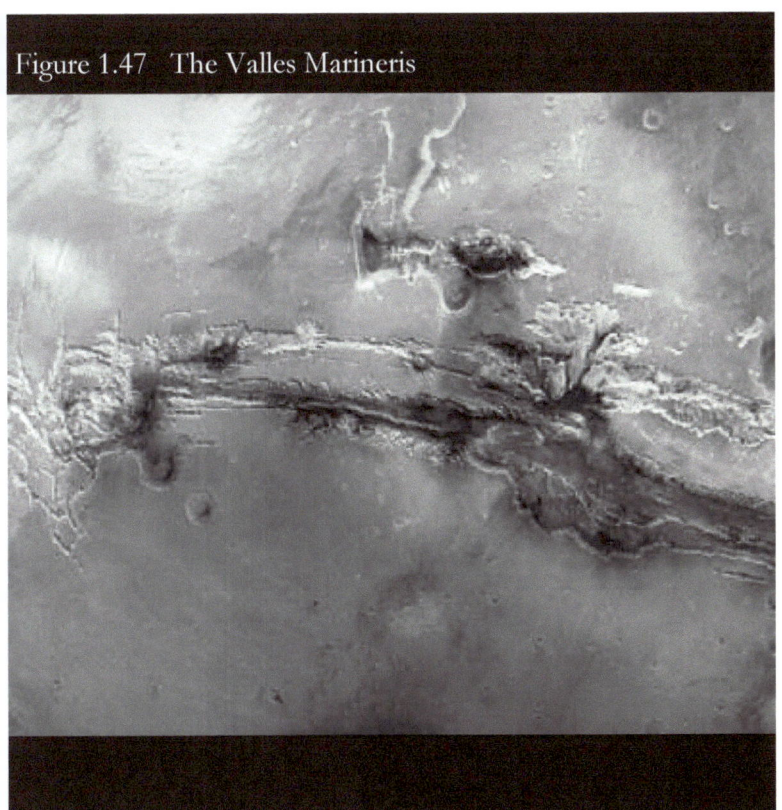
Figure 1.47   The Valles Marineris

1. (a) Using the formula, $K - 273° = °C$, transform the minimum and maximum temperatures found on the surface of Mars from Kelvin to Celsius.

   (b) Express the range of temperatures in both Kelvin and Celsius as inequalities, using "T" to represent temperature.

   (c) Express the range of elevation above the surface of Mars for the three volcanoes visible as dark spots. Provide two inequalities: one in miles and one in kilometers.

   (d) Express the depth of the Valles Marineris as a signed number.

figure 1.47 shows an artists rendering of the Pathfinder space vehicle entering the Martian atmosphere as it descends to the surface of the planet. The early goals for Pathfinder were simple. NASA wanted to demonstrate its newest technology to deliver a lander equipped with instrumentation along with a free-ranging robotic rover to the planet's surface. Pathfinder met this goal and also provided an unprecedented amount of data. Engineers designed an innovative method whereby Pathfinder entered the Martian atmosphere with a parachute to dampen the effects of gravity. In addition, Pathfinder was equipped with a massive system of airbags to cushion the impact.

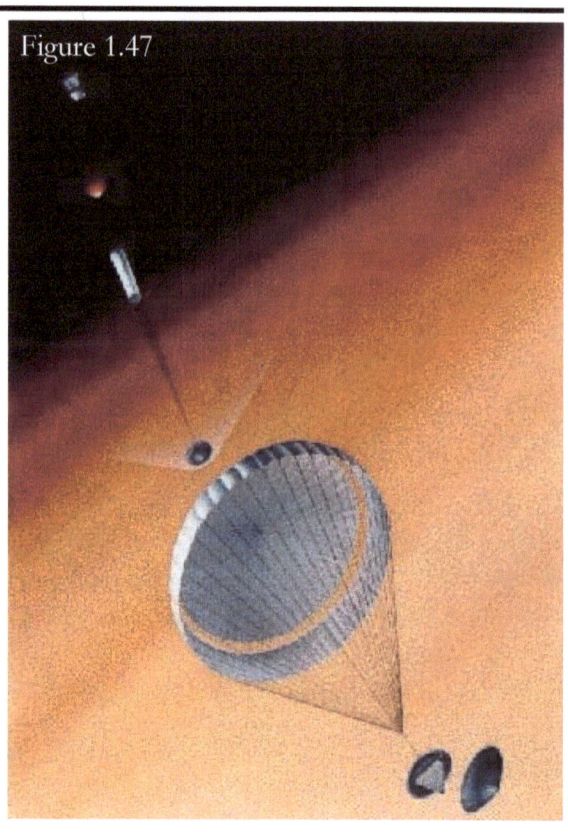

Figure 1.47

The parachute was able to dampen the effects of gravity by provided support slowing the craft as it was pulled to the surface. Unassisted, Pathfinder would have been pulled downward with the full force of Martian gravity, $g = -4$ m/sec², but instead, was pulled toward the surface with $g < -4$.

2. Complete the table by supplying the requested calculations. We are assuming that the Pathfinder space vehicle entered the Martian atmosphere and was accelerated by Mars gravitational constant, $g = -4$ for the times noted.

Then, you are asked to also supply the same calculations assuming that the parachute dampened the acceleration rate for the value indicated, $g = -2$. Obviously changes between an unassisted fall through the Martian atmosphere and the fall actually experienced by the lander with the aid of a parachute would occur. You are asked to calculate these changes.

Table 1.17

Time (sec)	Speed: $g = -4$ (m/sec²)	Absolute value	Speed: $g = -2$ (m/sec²)	Absolute Value	Changes in Speed between $g = -4$ and $g = -2$
t = 10					
t = 14					
t = 16					
t = 20					
t = 25					
t = 30					
t = 35					

✎ Assignment #23
Page 3 of 3   ☑ Investigation 1.7   Name

3. Write an inequality expression describing the range of speeds corresponding to the time ranges shown in the table below. (Recall example provided in Assignment #22).

Table 1.18

Time Range (sec)	Speed (Inequality) Expression (g = -2)	Speed (Inequality) Expression (g = -4)
$10 \le t \le 14$		
$14 \le t \le 16$		
$16 \le t \le 20$		
$20 \le t \le 25$		
$25 \le t \le 30$		
$30 \le t \le 35$		

Challenge Questions! Instructions: For the problems 4—12 below, insert < , > or = between each pair of numbers to make a true statement.

4. $-3 \quad\quad -5$

5. $-17 \quad\quad -6$

6. $|-9| \quad\quad |-15|$

7. $|-8| \quad\quad |-4|$

8. $-33 \quad\quad |-33|$

9. $-|-10| \quad\quad |-10|$

10. $-|-4| \quad\quad -|-8|$

11. $-|-4-5| \quad\quad -|-3-8|$

12. $-|-5+6| \quad\quad -|-5-6|$

For problems 13—24 below, add, subtract, or multiply as indicated:

13. $-7 + 12 =$

14. $(-7)(12) =$

15. $-7 - 12 =$

16. $(-9)(10) =$

17. $-9 - 10 =$

18. $-9 + 10 =$

19. $26 - (-25) =$

20. $-10 - (-20) =$

21. $(-20)(10) =$

22. $-20 + 10 =$

23. $(-10) + (-20) =$

24. $38 - 44 =$

Math Toolbox

Investigation 1.8: Combining Concepts

Workshop 1: Signed Numbers

Figure 1.48

What are the "bends"?

Normally associated with SCUBA diving, "the bends" occur if a SCUBA diver stays under water at a depth > 30 meters about 100 feet and fails to ascend slowly.

For a certain period of time, some amount of **nitrogen** from the air will dissolve in the water along with his or her body. If the diver were to swim quickly to the surface, it is just like uncorking a bottle of soda -- the gas is released. This can cause a very painful condition, and is sometimes fatal. To prevent "the bends" a diver is required to ascend slowly, carrying out *decompression stops* on his/her way to the surface so that the gas can come out of the blood solution slowly.

If the diver does rise too fast, the only cure for "the bends" is to enter a pressurized chamber in which the air pressure matches the pressure experienced by the diver at the maximum diving depth. In this way, the pressure is released slowly.

If the diver decompresses properly, remains at "recreational depths" (less than 100 feet or so), and is careful about the air supply, the dangers can be largely eliminated. Proper training, good equipment and careful execution are the keys to safe diving.

The Ocean's Surface

Step 1 — $1 \bullet (-8) = -8$

Step 2 — $2 \bullet (-8) = -16$

Step 3 — $3 \bullet (-8) = -24$

Step 4 — $4 \bullet (-8) = -32$

Step 5 — $5 \bullet (-8) = -40$

Example 1: Introduction to Algebra Patterns

A deep sea diver plans a deep water dive, intending to move down from the surface in short steps of 8 meters each. Suppose the diver moves down using a total of 5 such short steps shown at the left. Represent his movement downward as a product of integers and find the final depth.

How many variables are changing? Which ones?

- Let each **step** be represented by the variable, X.
- Let the **depth** be represented by the variable, Y.

If so, the movement downward can be represented by the following equation:

$$Y = (-8) \bullet X$$

100

Math Toolbox

Investigation 1.8: Combining Concepts

Workshop 1: Signed Numbers

Assume the deep sea diver continues his downward trek. Using the equation generated for the stepwise descent below the ocean's surface, find the depth for each of the short steps planned by the diver and enter them in the table below. Be sure to give your answer using the correct sign to represent "depth below the ocean's surface".

$$Y = (-8) \bullet X$$

Table 1.19

Step # (X)	Depth, m (Y)	Step # (X)	Depth, m (Y)
6	$Y = (-8) \bullet 6 = -48$	10	
7		11	$Y = (-8) \bullet 11 = -88$
8		12	
9	$Y = (-8) \bullet 9 = -72$	13	

What is the change in depth for each step?

Is it constant? Why or why not?

What relationship exists between the points in Figure 1.49 below?

Do you see a straight line?

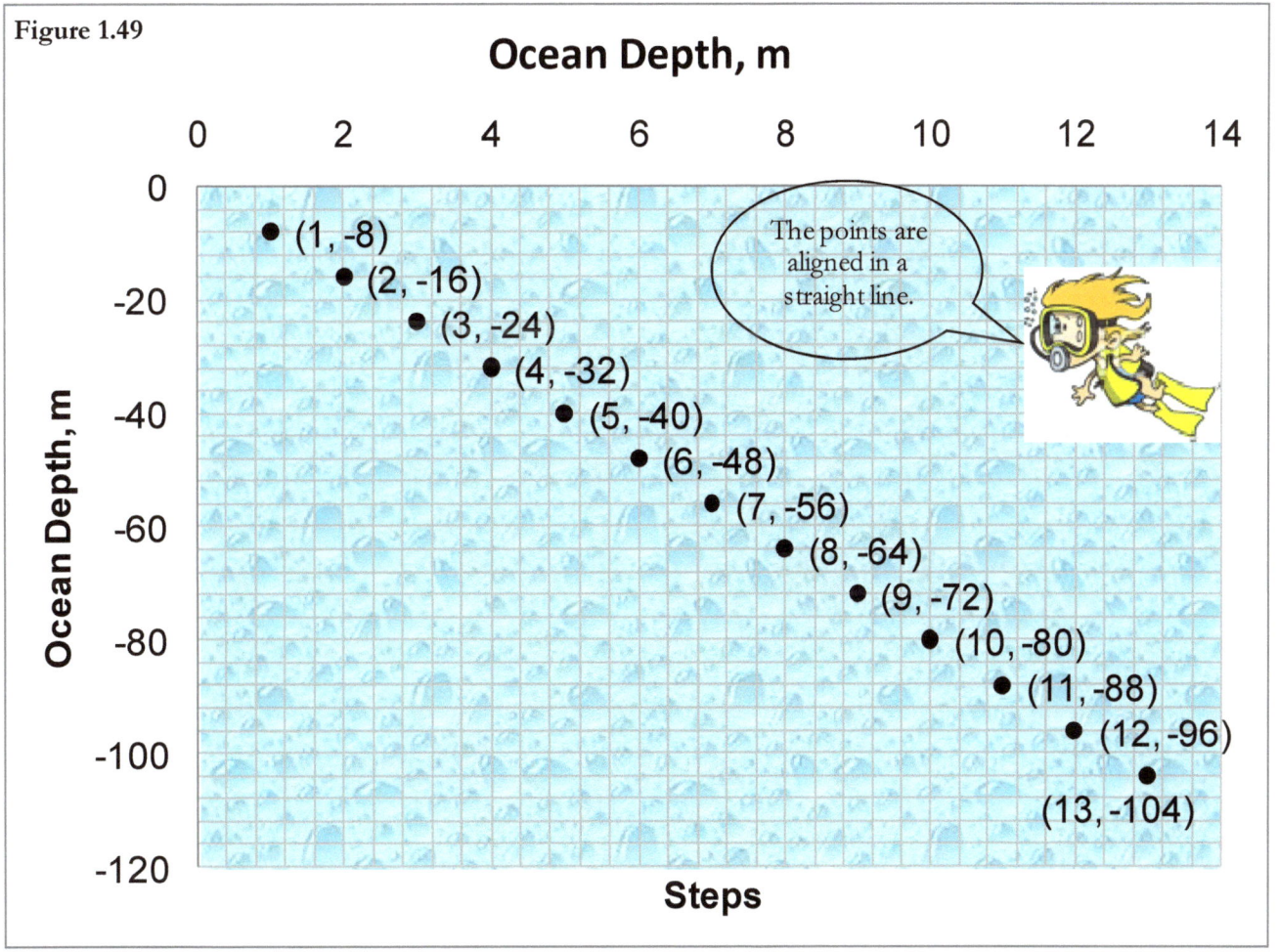

Figure 1.49

Math Toolbox

Workshop 1: Signed Numbers

Investigation 1.8: Combining Concepts

Example 2: Avoiding the "bends"

After reaching the diving depth of –104 meters, it becomes time for the deep sea diver to ascend to the surface. To be very sure, he doesn't become disabled by the "bends", his ascent will also be made using small, intermediate steps of four meters each. Remember that the negative sign indicated direction and because on the way down, the diver stopped every 8 meters, this was represented as –8.

On the way up, the diver will stop every 4 meters to avoid the bends. What sign will convey movement "up" rather than "movement down"? If you said, +, then you are correct.

- There are still two values that change: The step number (X), and the diver's depth (Y).
- Additionally, though, we must add the increased depth attained at each step to the starting value of –104 meters.

Step 1: At –104 meters, the diver moves upward 4 meters. At what depth is the diver?

$$Y = -104 + (4 \bullet 1)$$
$$Y = -100$$

Step 2: At –100 meters, the diver moves upward another 4 meters. At what depth is the diver?

$$Y = -104 + (4 \bullet 2)$$
$$Y = -104 + 8$$
$$Y = -96$$

Step 3: At –96 meters, the diver moves upward another 4 meters. At what depth is the diver?

$$Y = -104 + (4 \bullet 3)$$
$$Y = -104 + 12$$
$$Y = -92$$

Step 4: At –92 meters, the diver moves upward another 4 meters. At what depth is the diver?

$$Y = -104 + (4 \bullet 4)$$
$$Y = -104 + 16$$
$$Y = -88$$

Order of Operations
1. Do all operations within grouping symbols such as parenthesis or brackets.
2. Multiply or divide from left to right.
3. Add or subtract in order from left to right.

Math Toolbox

Investigation 1.8: Combining Concepts

Workshop 1: Signed Numbers

Step 5: At −88 meters, the diver moves upward another 4 meters. At what depth is the diver?

$Y = -104 + (4 \bullet 5)$

$Y = -104 + 16$

$Y = -84$

In Investigation 1.5 you were introduced to the idea that patterns exist in algebra. Specifically we referred to algebra, more than any other mathematical topic, is a study of patterns and this diving example illustrates this idea. When a pattern exists to describe, in this case, a predictable change in the diver's depth, we assign variables to simplify and understand the changing process: the step number (X), and the diver's depth (Y).

We now construct a general mathematical equation to find the diver's depth for any step number:

$$Y = -104 + (4X)$$

This equation acts as a formula allowing the diver's depth to be calculated by plugging in the particular step number. Use this equation to complete the table, then answer the Knowledge Check questions.

Table 1.20

Step #	Diver's Depth, (m)	Step #	Diver's Depth (m)	Step #	Diver's Depth (m)	Step #	Diver's Depth (m)	Step #	Diver's Depth (m)
1	-100	6	-80	11		16		21	-20
2	-96	7		12		17	-36	22	
3	-92	8		13	-52	18		23	
4	-88	9		14		19		24	
5	-84	10	-64	15		20	-24	25	

✓ **Knowledge Check:**

1. How many steps will the diver require to reach the ocean's surface?
2. Why is the number (-104) necessary in the equation? What does this number represent?
3. What does the number, "4" mean in this equation? Why is it necessary?
4. What does "X" represent in this equation?
5. What does "Y" represent in this equation?

Investigation 1.8: Combining Concepts

Workshop 1: Signed Numbers

Figure 1.50 is a graph of the stepwise change in the diver's depth below the surface.

✓ Can you identify the step and depth for each of the diver positions shown?

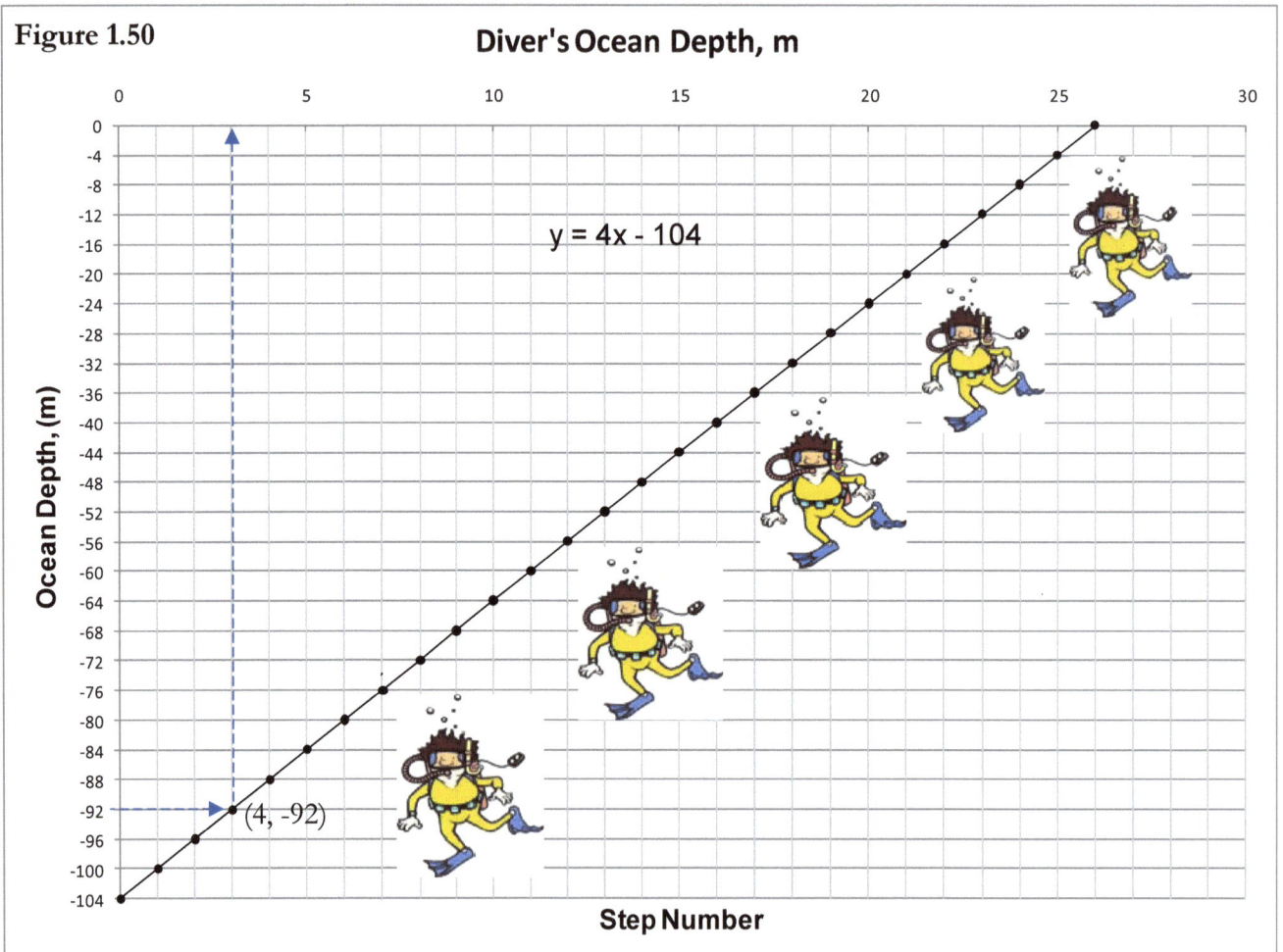

Figure 1.50

Write each (step #, Depth) combinations as an ordered pair. Using the values shown, at step 4, graph indicates the diver is at a depth of –92 meters below the surface. We write this—ordered pair—using parenthesis. The number graphed along the horizontal axis, **Step Number**, is listed first.

✓ The example shown on the graph, (4, -92) is interpreted this way:
■ At Step #4, the diver is located –92 meters below the surface of the ocean.

Assignment #24
Investigation 1.8

Remember Wanda? Seems her "widget" business has done well, however she had to borrow $2,000 for operating expenses in order to expand her business. Because this is money borrowed, Wanda expresses the money as a "debt", (-$2,000). In addition, the bank has agreed to a payment plan where each month Wanda pays $100. Your goal for this assignment is to complete the table below that establishes a pattern using a starting value of $-2,000.

Once you calculate the amount Wanda still owes after each of 7 monthly payments, look for the pattern and let X = the # of monthly payments, and Y = the amount Wanda still owes.

💲 Purpose an equation that will allow Wanda to calculate the amount of money still owed and use it to determine how many months it will take for Wanda to pay off this debt.

Table 1.21

# Months, (X)	Wanda's Debt, $ (Y)	# Months (Y)	Wanda's Debt, $ (Y)
0	-2000	4	
1		5	
2	-1800	6	-1400
3		7	

My Equation is:

Table 1.22

Month#	Wanda's Debt (Y)	Month #	Wanda's Debt (Y)	Month #	Wanda's Debt (Y)
0	-2,000	7		14	
1		8		15	
2	-1,800	9		16	
3		10		17	
4		11		18	
5		12		19	
6	-1400	13		20	

✎ Assignment #24
Page 2 of 2   ☑ **Investigation 1.8**   Name

Use Figure 1.51. The upper left hand corner of each box around *Wanda* is attached to a point. For each point, identify both the Month Number and Wanda's Debt. Then, write your answer as an ordered pair in the table provided.

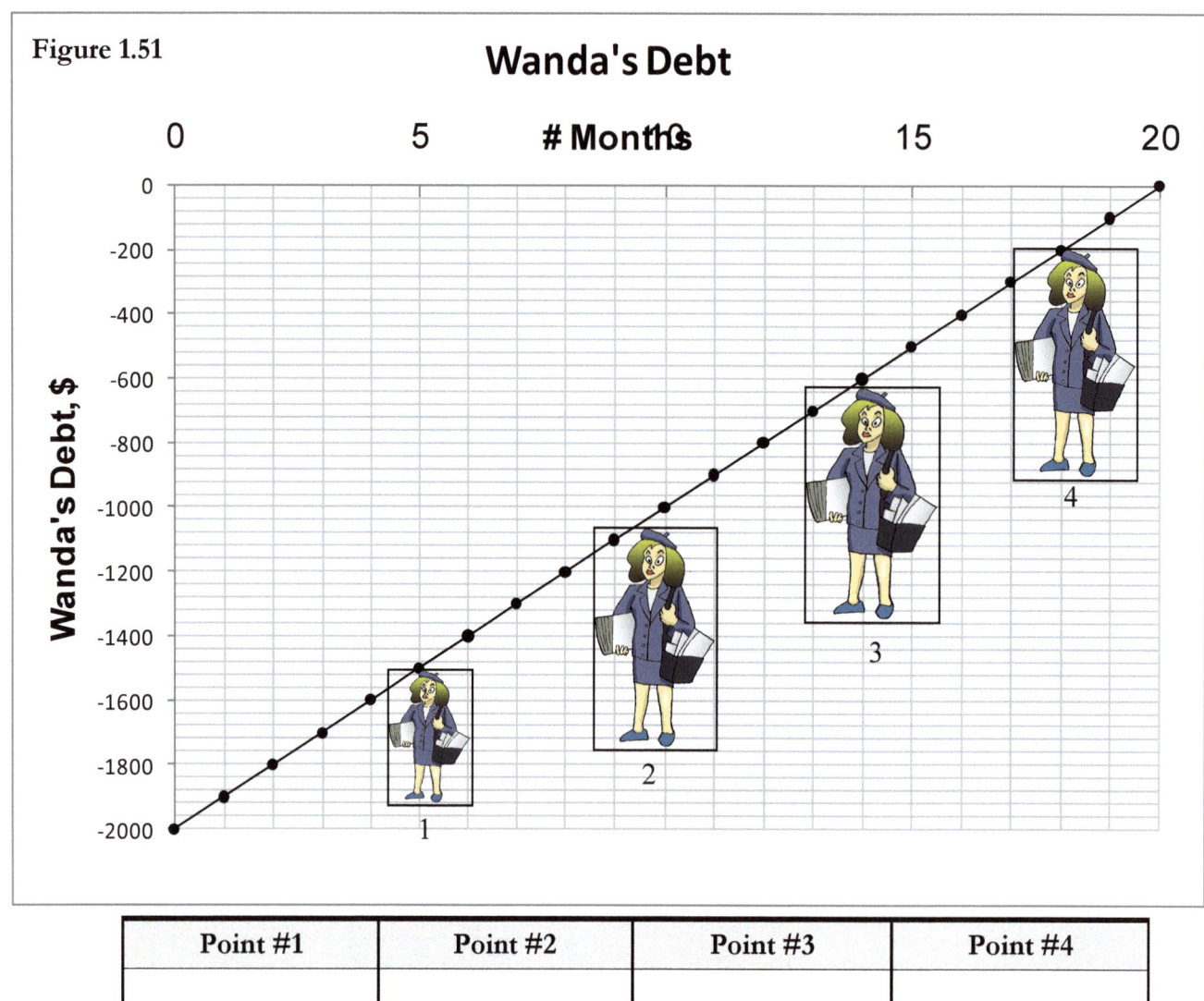

Point #1	Point #2	Point #3	Point #4

1. Why is the number −2000 important in your equation? What does it mean?

2. Why is the number 100 important in your equation? What does it mean?

3. What does "X" represent in your equation?

4. What does "Y" represent in your equation?

Assignment #25
Investigation 1.8

Saturn, you recall, is the second largest planet in the solar system and according to the official NASA website, was named by the ancient Greeks after the god of farming and time.

The rings of Saturn were first observed by Galileo, although he didn't know what they were, and in 1655 Christiaan Huygens discovered they were "rings".
Named in his honor by the European Space Agency, the Huygens space probe was sent to Titan, Saturn's largest moon, to investigate the its surface. The Huygen's probe hitched a ride on NASA's Cassini Spacecraft and was released into freefall on December 24, 2004. Huygens landed on Titan's surface January 14, 2005.

Figure 1.52

This image of Saturn was taken by Voyager I in 1980.
Courtesy of NASA

Figure 1.53

Huygens fell through Titan's atmosphere and this artists conception of the event shows the probe hanging from its parachute, with Saturn in the background.

Courtesy, *European Space Agency*

A heat shield protected Huygens as it entered the atmosphere, being pulled downward by Titan's low gravitational constant, about -2 m/sec^2.

Then, a series of parachutes deployed to slow the probe down gently lowering it through Titan's dense atmosphere. As it descended on a 2.5 hour ride, Huygens took measurements, sending important data back to NASA, finally landing on the surface with a frigid temperature of –178^0 C (-289^0 F). This temperatures corresponds to 95 on the Kelvin Scale.

Figure 1.54 Colored image released from the landing site.

The Huygens lander continued to send data back to NASA for a full 90 minutes after landing.

Although it isn't known for sure, let's assume that the series of parachutes deployed by Huygens cut the lander's descent rate in half. Use the graph on the following page to collect data about the Huygens lander as it descended to its planetary resting place shown in Figure 1.54.

Assignment #25
Page 2 of 3 — Investigation 1.8

Figure 1.55 indicates the descent speed for the Huygens probe as it descended toward Titan's surface. The graph below indicates several things: (i) the descent toward the planet's surface under the influence of Titan's gravitational field ($g = -2$ m/sec^2) (ii) the point at which a parachute was deployed, slowing Huygen's acceleration downward from -2 to -1 m/s^2 (iii) A running total of Huygens changing speed as it descended. Use this graph to complete the table on the following page by providing the missing values.

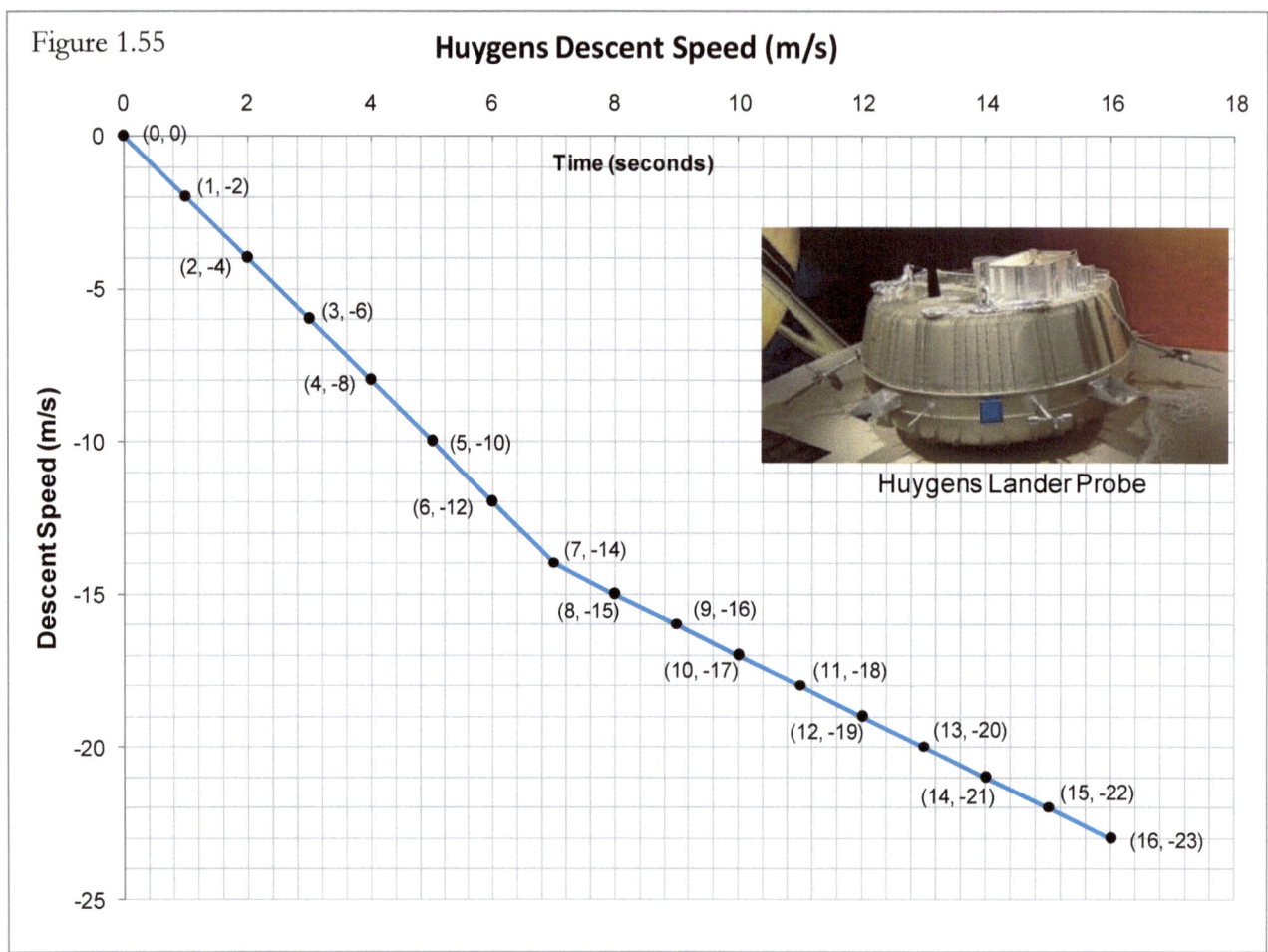

Figure 1.55 — Huygens Descent Speed (m/s)

A few things for you to consider while studying the graph of Huygens descent to the surface of Titan:

1. Locate the point on the graph indicating parachute deployment. How does the graph change at this point?
2. What is the change in speed from t = 5 seconds to t = 6 seconds?
3. What is the change in speed from t = 6 seconds to t = 7 seconds?
4. What is the change in speed from t = 7 seconds to t = 8 seconds
5. What is the change in speed from t = 8 seconds to t = 9 seconds?
6. What is the speed after t = 16 seconds?
7. What is the speed at t = 0 seconds?
8. Using t = Time, g = gravity, can you determine the equation used to generate the points on this graph?

✎ Assignment #25
Page 3 of 3   ☑ **Investigation 1.8**   Name

Using t = time, g = gravity, can you determine the 2 equations used to generate the points on Figure 1.55?

Time (sec)	Speed, S (m/s)	$|S|$ m/s	ΔS m/s	$\Delta |S|$ m/s
1	-2	+2	-2 - 0 = -2	+2
2	-4	+4	-4 - (-2) = -4 + (+2) = -2	+2
3	-6	+6	-6 - (-4) = -6 + (+4) = -2	+2
4				
5				
6				
7				
8				
9				
10				
11				
12				
13				
14				
15				
16				

Explain your answer to the following questions using a complete sentence paying attention to grammar, punctuation, and spelling.

1. What characteristic does "S" represent about the Huygens lander?

2. What does the absolute value of Huygens speed, S, indicate about the lander's descent?

3. What does ΔS indicate about the Huygens probe descent to the surface of Titan?

4. Interpret the constant **change in** Huygen's descent speed from -2 to -1 m/sec. What caused this?

5. Except for the change noted in Question #4 above, why is the speed relatively constant? What causes this?

www.ingramcontent.com/pod-product-compliance
Lightning Source LLC
Chambersburg PA
CBHW050726180526
45159CB00003B/1140